# Eukaryotes at the Subcellular Level

METHODS IN MOLECULAR BIOLOGY

Edited by

ALLEN I. LASKIN
*ESSO Research and Engineering Company*
*Linden, New Jersey*

JEROLD A. LAST
*Harvard University*
*Cambridge, Massachusetts*

# Eukaryotes at the Subcellular Level

DEVELOPMENT AND DIFFERENTIATION

Edited by Jerold Last

Harvard University
Cambridge, Massachusetts

MARCEL DEKKER, INC. New York and Basel

MARCEL DEKKER, INC.

270 Madison Avenue, New York, New York 10016

LIBRARY OF CONGRESS CATALOG CARD NUMBER: 75-42541

ISBN: 0-8247-6336-X

Current Printing (last digit):
10 9 8 7 6 5 4 3 2 1

PRINTED IN THE UNITED STATES OF AMERICA

PREFACE

This volume was initially conceived as a collection of methods, each applied to a completely different system, useful in the study of development and differentiation in higher organisms. It rapidly became something else: a compendium of methodology for the isolation, separation, analysis, and characterization of eukaryotic messenger RNA molecules. This seems to be the case because that is where the field presently is, and where it is likely to remain until a new generation of technology becomes available. The limiting factor in studies of differentiation by molecular biologists at present seems to be the ability to isolate a given mRNA in purity adequate to serve either as a template for reverse transcriptase to synthesize its complementary DNA or as an $^{125}$I-labeled probe itself. With labeled mRNA and/or cDNA one can determine the concentrations of mRNAs and of genes in various types of cells under a wide variety of conditions.

The specific mRNAs discussed in this volume include those characteristic of highly differentiated cells (erythrocytes, muscles, silkworm chorion, differentiating slime molds) and of hormonally responsive cells (oviduct).

As yet, specific mRNAs are most easily isolated from cells or tissues wherein their abundance is relatively high; the growing popularity of immuno-precipitation of specific polysomes as a first step in mRNA purification suggests that this technical limitation will soon be overcome and mRNAs for enzymes and other nonstructural proteins will also be isolated. Given that mRNA isolation plays such a dominant role in the current experiments of so many investigators, this book attempts to strike a balance between different systems in which the extant methodology has been developed and newer systems wherein the discoveries of the future may lie.

Jerold Last

# CONTRIBUTORS TO THIS VOLUME

JAMES F. COLLINS, National Institutes of Health, Bethesda, Maryland

HILDUR V. COLOT, The Biological Laboratories, Harvard University, Cambridge, Massachusetts

MORTON J. COWAN, National Institutes of Health, Bethesda, Maryland

RONALD G. CRYSTAL, National Institutes of Health, Bethesda, Maryland

ARGIRIS EFSTRATIADIS, The Biological Laboratories, Harvard University, Cambridge, Massachusetts

ALLAN JACOBSON, Department of Microbiology, University of Massachusetts Medical School, Worcester, Massachusetts

FOTIS C. KAFATOS, The Biological Laboratories, Harvard University, Cambridge, Massachusetts

JOANNE KAO, The Biological Laboratories, Harvard University, Cambridge, Massachusetts

ANTHONY R. MEANS, Department of Cell Biology, Baylor College of Medicine, Houston, Texas

BERT W. O'MALLEY, Department of Cell Biology, Baylor College of Medicine, Houston, Texas

KAY S. PINE, The Biological Laboratories, Harvard University, Cambridge, Massachusetts

SCOTT PORTNOFF, The Biological Laboratories, Harvard University, Cambridge, Massachusetts

JEFFREY M. ROSEN, Department of Cell Biology, Baylor College of Medicine, Houston, Texas

SATYAPRIYA SARKAR, Boston Biomedical Research Institute Boston, Massachusetts

ALLEN J. TOBIN, Biology Department, University of California, Los Angeles, California

PATRICK J. WILLIAMSON, National Institutes of Health, Bethesda, Maryland

SAVIO L. C. WOO, Department of Cell Biology, Baylor College of Medicine, Houston, Texas

NICHOLAS N. ZAGRIS, The Biological Laboratories, Harvard University, Cambridge, Massachusetts

NANCY ZARIN, The Biological Laboratories, Harvard University, Cambridge, Massachusetts

CONTENTS

Chapter 1

# THE CHORION OF INSECTS: TECHNIQUES AND PERSPECTIVES

A. Efstratiadis and F. C. Kafatos
The Biological Laboratories
Harvard University
Cambridge, Massachusetts

# I. INTRODUCTION

## A. General

A key aspect of cellular differentiation is programmed synthesis of specific proteins. Many cell types show characteristic developmental kinetics at the level of protein synthesis: the rates of synthesis of each of several specific proteins change with time, in reproducible temporal relationship to each other. In other words, the expression of specific genes, via proteins, is sequentially regulated, and we can talk about differentiation as a program of differential gene expression (Fig. 1). Typically, this program is initiated by an external stimulus (such as a hormone or an inducer) but then proceeds autonomously. A major goal of developmental biology is to understand the intracellular regulatory mechanisms that ensure timing and quantitative control of sequential gene expression [1].

For reasons of methodology, it is desirable to use systems in which the specific proteins are synthesized at high rates. The strategy of biochemists working in this field is "from specific proteins to specific genes." One first wants to describe the system thoroughly, by obtaining quantitative measurements of the rates of protein synthesis over time. The next step is to isolate and identify the corresponding mRNAs. One can then devise assays that will reveal directly the program of synthesis and accumulation of these mRNAs. By correlating mRNA metabolism with the program of specific protein synthesis, one might hope to evaluate the relative importance of various steps in the flow of genetic information (transcriptional and post-transcriptional). Ultimately, one can hope to use the mRNA as a handle to identify and purify the corresponding genes, together with their associated sequences. On the reasonable assumption that transcriptional regulation will prove to be both important and related to sequence organization, an obvious goal is to obtain large quantities of this specific DNA for use in detailed studies of sequence organization and of cell-free transcription. Artificial DNA amplification via phages and plasmids [2,3] will soon be a routine procedure, one hopes, with the proper safeguards[4].

A complementary strategy entails the use of genetics. One would like to collect mutants in many of the cistrons regulating a developmental program of specific protein synthesis and then characterize them biochemically (in terms of changes in specific protein and mRNA metabolism), genetically (in terms of complementation, linkage, and dominance relationships), and physiologically ( in terms of responsiveness to signals such as hormones, and of timing of action as assayed by periods of thermosensitivity in the case of temperature-sensitive mutants). The use of mutants can provide rigorous proof of molecular regulatory models, as well as lead to novel models. This powerful approach has not as yet been applied seriously in studies of eukaryotic differentiation. In practical terms, it seems clear that it can only be attempted in species already well characterized genetically -- yeast and *Drosophila* for now, and probably *Dictyostelium* and the nematodes in the future. To facilitate biochemical analysis, once again it is desirable to choose systems in which specific proteins are synthesized sequentially in large amounts.

Neither specific DNA isolation nor the genetic approach will be discussed further in this chapter. The attractiveness of studies on the chorion system, however, is partly related to their ultimate extension in these directions.

## B. The Chorion System: From Specific Proteins to Specific Genes

The follicular epithelium of insects is a layer of cells surrounding the oocyte. As the end of oocyte maturation approaches, these cells become highly specialized for the production of a group of structural proteins that are secreted extracellularly and assemble to form the protective eggshell or chorion, surrounding the oocyte [1,5]. The biology of this system can be summarized briefly by reference to production of the chorion in the wild silk moths (family Saturniidae).

The mature silk moth chorion can be easily obtained by dissection from the egg (major diameter, ca. 2 mm), since it is an insoluble structure of considerable size (ca. 0.5 mg in *Antheraea polyphemus*). Each moth can yield more than 100 eggs, and thus gram quantities of purified chorion are available. The eggshell is dissolved when treated with a combination of a denaturant and a reducing agent (e.g., SDS or urea and mercaptoethanol or DTT). The dissolved proteins can be analyzed by electrophoresis on SDS-polyacrylamide gels and by isoelectric focusing (IF); approximately 15 and 40 components are resolved by these methods, respectively (Figs. 2 and 3). The components resolved on SDS gels are synthesized as distinct proteins and are not derived from each other or from larger precursors. Some of the components resolvable on IF gels are formed by post-translational modification (J. Regier, unpublished observations).

During the period of choriogenesis, the follicular cells devote virtually their entire protein synthetic capacity (>90%) to production of the chorion protein [5]. Moreover, the pattern of chorion protein synthesis changes dramatically and reproducibly with developmental stage. Based on the relative synthetic rates of proteins recognized on SDS gels, 18 stages of differentiation have been defined in *A. polyphemus* and shown to last approximately three hours each (Fig. 4). The developmental program of chorion protein synthesis is autonomous, in the sense that it can be implemented faithfully not only in vivo but also in short-term organ culture, in which individual follicles are maintained in a defined medium [6].

Description of this program of specific protein synthesis was facilitated by the high degree of specialization of the cells and by our ability to recognize the chorion proteins (Figs. 2 and 3; Table 1), by virtue of their characteristic (and unusually low) molecular weight, their characteristic (and unusually acidic) isoelectric point, and their unusual amino acid composition (a high percentage of hydrophobic amino acids, especially glycine and alanine, and a high cysteine content). In addition, chorion-specific antibody is available. A number of chorion proteins have been purified to homogeneity, and their structure and similarity to each other are under intensive investigation. It is already clear that the proteins are closely related, but encoded by distinct genes (J. Regier, unpublished observations). Antibodies to individual proteins are now being generated.

A meaningful investigation of the regulation of this program depends on our ability to recognize and quantify the chorion messages. We have been able to purify and display the chorion messages on polyacrylamide gels [7]. The nature of the characteristic RNA bands as chorion messages has been confirmed by their ability to program the synthesis of immunochemically recognizable chorion proteins in the ascites and wheat germ cell-free systems [8], by a detailed correlation of the size of these RNAs and the size of the chorion proteins produced in different silk moth species [7], by their stage specificity and association with polysomes during the period of chorion

TABLE 1

Composition of Chorion in Some Insects (Residues per 100 Residues Recovered)

| | *A. polyphemus*[a] | *A. pernyi*[b] | *B. mori*[b] | *D. melanogaster*[a] |
|---|---|---|---|---|
| Glu-$NH_2$ | N[c] | 0.0 | 0.0 | 0.6 |
| Gal-$NH_2$ | N | 0.0 | 0.0 | 0.4 |
| Trp | 1.6 | 0.8 | 1.4 | 1.0 |
| Lys | 0.4 | 0.5 | 0.2 | 5.0 |
| His | 0.1 | 0.0 | 0.3 | 2.2 |
| Arg | 2.4 | 2.4 | 2.6 | 3.8 |
| Cys | 6.1 | 5.8 | 12.0 | 0.6 |
| Asx | 3.9 | 3.6 | 4.3 | 5.9 |
| Thr | 2.8 | 2.9 | 3.1 | 2.4 |
| Ser | 3.7 | 3.7 | 3.5 | 8.4 |
| Glx | 4.7 | 4.4 | 3.7 | 10.1 |
| Pro | 3.0 | 3.8 | 4.0 | 10.9 |
| Gly | 32.9 | 32.6 | 36.0 | 15.7 |
| Ala | 11.9 | 12.7 | 7.0 | 14.7 |
| Val | 6.6 | 6.6 | 5.9 | 6.1 |
| Met | 0.4 | 0.3 | 0.1 | 0.5 |
| Ile | 3.8 | 3.9 | 2.8 | 3.4 |
| Leu | 7.7 | 7.2 | 5.1 | 4.9 |
| Tyr | 6.6 | 7.4 | 6.3 | 3.8 |
| Phe | 1.5 | 1.4 | 1.8 | 1.2 |
| Amide[d] | N[c] | 42.5 | 42.5 | N[c] |
| Phosphate | N[c] | 0.0 | 0.0 | N[c] |

[a]From Ref. 1. Data are for soluble proteins only (*A. polyphemus*, proteins soluble in urea + DTT) and for Zones II + III only (*D. melanogaster*, chorion proper).
[b]From Ref. 13.
[c]Not determined.
[d]Expressed as molar percent of Asx and Glx.

formation [9], by their poly(A) content [10], and by their base composition (high GC, as predicted from the amino acid composition [1]), So far the messages have been fractionated with limited resolution (Fig. 5), into two main zones and one minor zone (zones 1 and 2; zone 3 is usually evident as a shoulder of zone 2). Our working assumption is that they correspond to messages for the three protein size classes, A, B, and C, respectively.

Synthesis and accumulation of the three chorion mRNA classes (at least in their mature, polyadenylated, and translatable form) proceed in parallel with chorion protein synthesis [1,9]. Detailed correlation of the developmental kinetics of synthesis of a single protein and its mRNA has not been undertaken, since we cannot as yet recognize individual mRNA species. The [$^3$H]cDNA probe has been synthesized, to be used in studies of the metabolism of mRNA precursors and in the study and isolation of chorion genes.

We have recently begun a parallel study of the chorion system in Drosophila, as well [1, 11]. In its broad outline, the system is quite similar to that in silk moths, although the proteins are chemically different (Table 1). The size of the eggshell is almost three orders of magnitude less than that of silk moths, but mass isolation procedures have been devised that, within a few hours, yield milligram quantities of purified, dissolved chorion proteins and of living follicle suitable for biosynthetic studies. The attractiveness of this system is that chorion mutants can be easily generated; through the study of female-sterile mutants, 11 cistrons that affect the chorion have been tentatively defined on the X chromosome (D. Mohler, private communication).

In the remainder of this chapter, we will discuss some methods to be used in going "from specific proteins to specific genes," with special reference to our work on the silk moth chorion system.

## II. FRACTIONATION OF SPECIFIC PROTEINS

A detailed treatment of this topic is beyond the scope of this chapter. For the sake of completeness, we will refer to the various properties generally used as the basis of protein fractionation, and will then concentrate on charge, size, and shape, which are the most generally applicable of these properties.

### A. Biologic Compartmentalization

Certain proteins can be partially purified on the basis of their association with particular cell constituents. Examples would be the histones and other chromosomal proteins, membrane proteins, etc. Extracellular proteins may be purified by removal of the cells, and this is in fact how we obtain pure chorion. In the silk moth, the chorion is large enough to be obtained by dissection, whereas in *Drosophila*, the oocyte is separated from the chorion by osmotic shock, and the chorion is then recovered by differential sedimentation [11, 12].

### B. Specific Affinity

Proteins can be purified by taking advantage of their specific affinity for substrates or inhibitors (e.g., enzymes) or particular macromolecules (e.g., DNA-binding proteins). Once a protein has been purified and used as an immunogen, the antibodies it elicits can in turn be used as affinity probes for subsequent recovery of the protein from complex mixtures. In our work, immunoprecipitation has been used mostly to purify radioactive polypeptides from cell-free translation mixtures; procedures will be described in the section on cell-free translation.

### C. Solubility

Differential extraction and differential precipitation are very effective fractionation methods, especially applicable to structural proteins, which are often unusual in terms of solubility. These methods are particularly attractive when applicable, because of their simplicity and ease of scaling up. Silk moth chorion can be purified as a "scleroprotein fraction;" it is completely insoluble in the absence of denaturing and reducing agents [13]. The extracellular *Drosophila* chorion can also be purified from follicular homogenates because it is insoluble in the absence of strong denaturants. Some high-cysteine proteins in the chorion of the silk moth, *Bombyx mori*, remain insoluble in low concentrations of thiol reagents and can thus be purified from the more soluble,lower -cysteine chorion components. Differential solubility is the basis of the first step in purification of specific silk moth chorion proteins (methods developed by J. Regier; see Ref. 1). Preparation of protein derivatives can be very useful in fractionation. Thus, we routinely use the carbamidomethylated derivatives of silk moth chorion proteins (prepared by a reduced protein-iodoacetamide reaction), since they are more soluble than the unmodified proteins under certain conditions (as well as being resistant to oxidation). The most insoluble unmodified chorion proteins (some Bs and Cs) can be made relatively stable by aminoethylation (reaction with ethylenimine), which introduces additional charges.

### D. Charge, Size, and Shape

Charge, on the one hand, and size and shape on the other, are the most general bases for protein fractionation schemes. We treat them together, because they are encountered in a whole spectrum of procedures, ranging from "pure charge" to "pure size," as follows:

-Isoelectric focusing
-Ion exchange chromatography
-Liquid electrophoresis
-Gel electrophoresis
-Gel chromatography
-SDS gel electrophoresis
-Centrifugation (including gradient sedimentation)

These methods also vary in the amount of protein that can be fractionated and in their resolution. Chromatographic and centrifugation methods have a high capacity but a relatively low resolution, whereas the opposite is true for most gel electrophoretic and gel isoelectric focusing methods. Isoelectric focusing in sucrose gradient-stabilized liquid columns is a relatively high resolution preparative method. It can be repeated, using progressively narrower ampholine ranges, to further fractionation. This method has been used extensively for the final purification of individual chorion proteins (J. Regier, unpublished; see Ref. 1).

Analytic fractionation can be maximized by a combination of the highest resolution techniques at the two ends of the spectrum. O'Farrell recently developed precise procedures for a two-dimensional analytic fractionation, using cylindrical gel isoelectric focusing as the first dimension and SDS slab polyacrylamide gel electrophoresis as the second; the reader is referred to his paper, since the details of the procedure are very important and cannot be summarized briefly [14]. We have used the same principle, excising individual bands from isoelectric focusing gels and subjecting them to SDS cylindrical polyacrylamide gel electrophoresis. The method is as follows (M. Nadel, unpublished):

Upon completion of the run, the IF gel is put into 10% TCA in the cold. Within 5 to 10 min the precipitated bands appear. The desired bands are immediately excised (using an oblique light source to increase contrast), dipped in buffer (0.33M Tris-HCl, pH 8.4), rinsed with buffer several times until neutralized, cut into fragments, and incubated overnight in a small volume of sample buffer. For complete recovery, both the extract and the gel fragments are loaded on the SDS gel. Alternatively, the neutralized fragment can be loaded directly.

The two major analytic methods, polyacrylamide gel isoelectric focusing and SDS polyacrylamide gel electrophoresis, will be discussed separately.

Isoelectric Focusing

General. In addition to other publications [15, 16], a symposium volume summarizing theoretical and practical aspects of isoelectric focusing has recently appeared [17]. For our purposes, it is sufficient to recall that the nearly universally used LKB Ampholines are multicomponent mixtures of "carrier ampholytes," i.e., relatively low molecular weight ampholytic substances with a high degree of ionization and a high buffering capacity near their isoelectric point. (Carrier ampholytes can also be synthesized simply and cheaply [18].) These two properties, as well as the multicomponent nature of the mixture, are essential for the setting up of a relatively stable pH gradient under the influence of an electric field. Ampholines are mixtures of isomers and homologues of aliphatic polyamino polycarboxylic acid, obtained by coupling residues containing a carboxylic group (e.g., propionic acid) with mixtures of polyamines, of the general formula

$$(R)_2\text{-N-}(CH_2)_n\text{-N-}(R)_2$$

where R is H or $-(CH_2)_n\text{-N-}(R)_2$. The products obtained are generally

$$R\text{-N-}(CH_2)_n\text{-N-}(CH_2)_n\text{-COOH}$$

where R is H, $-(CH_2)_n\text{-N-}(R)_2$, or $-(CH_2)_n\text{-COOH}$, and $n < 5$ [16, 17]. The conditions of synthesis ensure the multicomponent nature of the product (several hundred species), which is particularly important for resolution, since two proteins of similar isoelectric point can be separated only if a carrier ampholyte of intermediate isoelectric point exists in the gradient [19]. Maximization of the degree of ionization and buffering capacity is ensured by the closeness of the various pK values of each molecule. When these mixtures are subjected to an electric field, each component migrates to form a band which assumes a position between (and, actually, partially overlapping) similar bands just higher and just lower in isoelectric point. As a result, a monotonically changing pH gradient is set up, at each point of which the pH is dictated by the isoelectric point of the ampholyte "focused" there. The overall shape of the pH gradient is dictated by the pIs, the concentrations, and the buffering properties of the ampholytes. With sufficiently numerous and sufficiently closely spaced components, the conductivity and the buffering capacity are high throughout, and the gradient remains stable for hours (although it gradually drifts, for not very well understood reasons).

When proteins are allowed to move in this gradient, they will migrate to their isoelectric point and then stop. It is necessary, of course, that the focused protein not overwhelm the pH gradient locally; this is ensured by a relatively small protein concentration (avoidance of overloading) and ΔpK values between the various charged groups of the proteins greater than the ΔpK values of the ampholytes (a condition generally fulfilled).

Once it focuses, each protein assumes a Gaussian distribution, with the maximum at the isoelectric point. The width of the distribution is determined by the balance between two opposing forces: electrophoresis toward the isoelectric point (which tends to make the band thinner) and diffusion (which tends to broaden it). Resolution can be evaluated from the minimum distance between the centers of two resolvable components, which is equal to three times the standard deviation of each distribution. It can be shown that this minimum resolving distance is given by

$$\Delta(\mathrm{pI}) = \sqrt{\frac{D(dpH/dx)}{E(-dm/dpH)}} \qquad (1)$$

where D is the diffusion coefficient of the proteins, dpH/dx the local slope of the pH gradient in the gel, E the electric field strength, and dm/dpH the mobility slope of the proteins in the pH gradient. The mobility slope and the diffusion coefficient are inherent properties of the proteins (of course, D is higher and the resolution lower in liquid columns; in gels, although D could be decreased by using high acrylamide concentrations, sieving effects must be avoided). Raising the field strength and flattening the pH gradient are thus the two methods for improving resolution. The limits are imposed, respectively, by overheating the gel and by the pI range covered by the proteins to be fractionated (as well as by the necessity of having enough species of ampholine in that pH range, so that no part of the gel is ampholine deficient; otherwise, as already stated, proteins cannot be resolved in that part of the gel and, in addition, conductivity drops and local overheating occurs). The maximum value of E that can be used depends on the provisions for cooling. Thin horizontal slab gels are ideal for isoelectric focusing work not only because they permit comparison of many samples and lend themselves to autoradiography, but also because they can be chilled efficiently by circulating water beneath their glass backing. A good design is the LKB 2117 Multiphor, which can also be duplicated in its essentials by a homemade version. With proper cooling, 800 V or higher can be applied to our gels (1.5 mm thick, 25.5 cm wide, 15 cm long in the direction of the current). It should be remembered that the voltage should be raised in steps, since conductivity is high (because of the movement of ampholines), until the pH gradient is established.

To bring the proteins near their final position in a particular type of IF gel, the product Vt (voltage times time) is important, and V and t can be adjusted accordingly for convenience, provided that the run is terminated at maximum voltage for approximately one hour, to sharpen the bands. Additional relationships that are useful to remember for selecting conditions for a run are:

$$Vt \propto (\text{gel length})^2$$
$$V \propto (\text{gel length})$$
$$V \propto (\text{gel width})$$

Procedures for Slab Gel Isoelectric Focusing of Silk Moth Proteins (M. Nadel and M. Goldsmith, unpublished).

Sample preparation. The chorions are dissolved within two hours at room temperature with shaking, in sample buffer containing 8M urea, 0.07M DTT, 0.05M Tris-HCl (pH 9.0). Usually we dissolve 20 Bombyx chorions (0.8 mg protein) in 100 μl of sample buffer. Convenient containers are microanalysis tubes (Werthemann and Co., Basel, Switzerland), but microfuge tubes can also be used. To reduce the chances of protein deamidation, the buffer can also be made to contain 1.5 mM lysine and 1.5 mM EDTA. Only highest purity (cyanate-free) urea can be used, and all solutions must be freshly made or stored frozen. Sample buffer is frozen, capped, in small aliquots used only once. Stock urea solutions can be deionized by shaking with distilled water-washed Rexyn I-300 resin (Fisher), and filtering, and should never be heated above room temperature.

To block sulfhydryl groups and reduce the chance of oxidation during the run, the dissolved proteins are carbamidomethylated by reaction with iodoacetamide. To the protein solution is added one-half volume of 0.36 or 0.42M iodoacetamide (Eastman, recrystallized from water) in 1.2M Tris-HCl (pH 8.4) (made just before use, and kept in the dark throughout; the iodoacetamide is in 1.2 x and 1.4 x excess, respectively, relative to total sulfhydryl, and the lower concentration may be preferable). The reaction is terminated after 15 min by the addition of an excess of mercaptoethanol (Eastman, 3 μl/100 μl final solution).

Choice of Ampholines. We have found useful the

following formulations of ampholines:

1. pH 2.5-8: 0.4% (pH 2.5-4) ampholytes
0.8% (pH 4-6) ampholytes
0.4% (pH 5-8) ampholytes
The basic and acidic electrode buffers are 0.5% (pH 5-8) ampholyte and 1M $H_3PO_4$, respectively.
2. pH 4-6: 2% (pH 4-6) ampholytes
The basic and acidic electrode buffers are 1M NaOH and 1M $H_3PO_4$, respectively.

It may be possible to use lower ampholyte concentrations, permitting substantial savings.

Gel Preparation (5% acrylamide, 0.2% bis, 6M urea). The highest purity chemicals are used, or, alternatively, chemicals are recrystallized as shown. Caution: acrylamide and bis are poisonous and are absorbed through the skin. Also, upon recrystallizing, separate the crystals from the mother liquor within three hours.

Acrylamide Stock Solution

5 g acrylamide (recrystallized from acetone, 1 X, chloroform 2 X, dry 5-6 days in dark; or Bio Rad, electrophoresis grade).
0.2 g bis (N, N'-methylene-bis-acrylamide, recrystallized from acetone 2 X; or Bio Rad, electrophoresis grade).
36 g urea (recrystallized from ethanol; or Schwarz-Mann, Ultrapure).
Bring to final volume of 95 ml with distilled $H_2O$ and filter.

To Make Gel

68 ml acrylamide stock solution.
3.6 ml ampholine solution (3.4 ml for pH 2.5-8).
180 µl 10% ammonium persulfate solution (Bio Rad, electrophoresis grade; or Fisher).
Degas.
Immediately add 15 µl TEMED (Eastman) and pour gel between two clean glass plates, through a cut in the rubber gasket separating them along their periphery (1/16" thick, or thinner).

Sample Application. The polymerized gel is placed horizontally on the cooling plate, and the top glass is removed. Filter paper strips (LKB strips for electrode solution) impregnated with the electrode buffers are placed at opposite ends, at positions where they will make contact with the platinum wires of the cover when the latter is placed in position.

Samples (5-15 μl) are spotted on strips of Whatman 3 MM paper (2.5 to 5 mm x 10 to 20 mm; if the sample is dilute, several such strips can be impregnated and stacked at the same location). To check whether the isoelectric pattern at the end of the run is at equilibrium and free of artifacts (e.g., precipitation bands), one can place duplicate samples at different postions between the electrodes. Normally we apply the voltage across the 15 cm dimension of the gel.

Running and Subsequent Processing of the Gel. Run on the cooling plate at 8-10° C for about five hours, bringing the voltage to 800 V during the first hour (while the current is maintained at $\leq$ 25 mA).

When the run is complete, float the gel into 10% TCA, change the TCA several times (overnight), stain with 0.03% Coomassie brilliant blue in 5:5:1 $H_2O$: methanol: acetic acid for three hours, and destain in 5% methanol - 7.5% HAc. At this stage, the gel should be photographed for permanent record. We place the gel, in an open dish containing destaining solution, on a light box and photograph it with 4 x 5 inch Ektapan (Kodak) negative, which is subsequently developed with D-11 developer. To enhance contrast, an orange filter can be used.

Drying, Autoradiography, or Fluorography. The gel is dried by the procedure of Maizels [20] and autoradiographed with Kodak RP-R54 Royal X-Omat film, held in a Kodak X-ray exposure holder. To increase sensitivity (for $^{14}C$ or $^{35}S$) or to make possible the detection of $^{3}H$, fluorography rather than autoradiography can be used [21]. The gel is impregnated with PPO and exposed at $-70^{o}$ C or below. Fluorography has a lower resolution than autoradiography, because of the long path length of photons as compared to β rays.

Comments on Isoelectric Focusing. This is a procedure with very high resolution, by which one can separate proteins differing in pI by only 0.02 pH units; single charge differences between proteins can usually be detected with ease. At the same time, the high resolution can reveal annoying artifactual microheterogeneity (as well as biologically interesting post-translational modifications). For reproducibility, the chosen protocols should be followed scrupulously, the behavior of the power supply and cooling water bath should be monitored, and fresh chemicals should be used.

Mature chorions of various silk moths yield approximately 40 bands in our IF slabs. Pulse-chase experiments reveal that some of these components are post-translationally modified, whereas others appear not to be (J. Regier, unpublished). It is not yet known whether more than one band obtained from a mature chorion is derived from any one primary translation product. The proteins resolved by SDS gel electrophoresis show no interconversion in pulse-chase experiments and are not derived from large precursors [1, 5]; this suggests that the post-translational chorion protein modifications are minor in terms of mass, and validates the use of protein synthetic profiles as evaluated on SDS gels (Fig. 4) in developmental studies.

## SDS-polyacrylamide Gel Electrophoresis

General. A good introduction to gel electrophoresis is given by Maurer [22] . Of the review articles, we have found that of Weber et al. [23] particularly useful.

In order to understand gel electrophoresis, it is necessary to keep in mind some aspects of free electrophoresis [22]. In response to an electric field of strength E, applied across a liquid column, charged particles within the column show a mobility, $m_0$, which is

$$m_0 = \frac{d}{t \cdot E} = \frac{Q}{f} \qquad (2)$$

where d is the distance traveled, t the time, Q the effective charge of the particle, and f the frictional resistance ($f = 6\pi r v \eta$, where r is the radius of the particle, v the velocity, and $\eta$ the viscosity of the medium). Because of the effect of the ionic strength, $\mu$, on the effective charge of the particle,

$$m_0 \propto \frac{1}{\sqrt{\mu}} \tag{3}$$

Free electrophoresis tends to separate macromolecules on the basis of their charge/mass ratio. Stronger size dependence can be introduced by performing electrophoresis in a gel, where the pores exert a sieving effect. Gels of variable porosity can be constructed by polymerizing appropriate mixtures of acrylamide and bisacrylamide (bis),

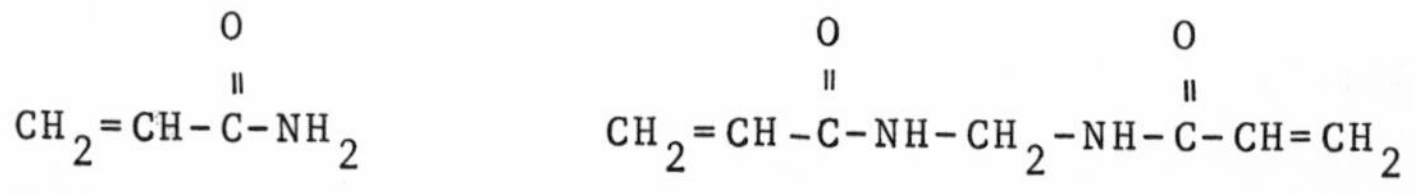

$$CH_2{=}CH{-}\overset{\overset{O}{\|}}{C}{-}NH_2 \qquad\qquad CH_2{=}CH{-}\overset{\overset{O}{\|}}{C}{-}NH{-}CH_2{-}NH{-}\overset{\overset{O}{\|}}{C}{-}CH{=}CH_2$$

Acrylamide N, N'-methylene-bis-acrylamide

thus constructing a three-dimensional polymer network in which bis forms the cross-links:

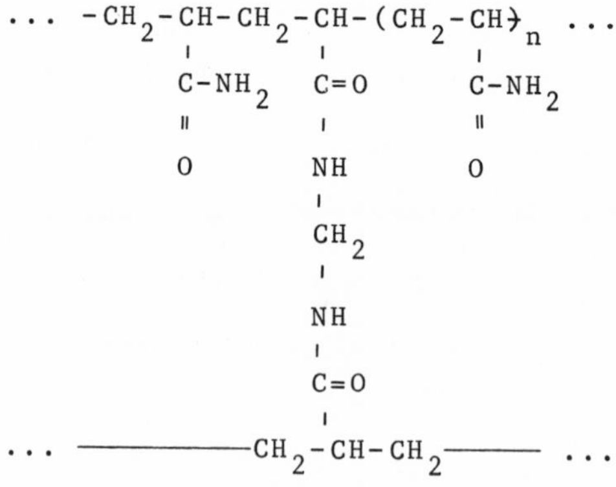

```
... -CH2-CH-CH2-CH-(CH2-CH)n ...
         |      |        |
       C-NH2   C=O     C-NH2
         ||     |        ||
         O      NH       O
                |
                CH2
                |
                NH
                |
                C=O
                |
... ---------CH2-CH-CH2--------- ...
```

The porosity can be specified by the percent total gel concentration, T, and percent cross-linking, C

$$T = a + b; \; C = \frac{b}{a + b} \times 100 \qquad (4)$$

where a and b are the concentrations of acrylamide and bis, respectively, in the polymerizing solution, in g/100 ml.

In general, from studies of protein mobility (m) in polyacrylamide gels, it is observed that

$$\ell og\ m = \ell og\ m_0 - KT \qquad (5)$$

where K is a size dependent factor or retardation coefficient. For molecules of similar shape and partial specific volume, K is given by

$$K = k_c MW + A \qquad (6)$$

where A is a constant and $k_c$ a cross-linkage coefficient, dependent on the value of C. From Eq. (5) and (6), it can be seen that the general relationship between molecular weight and mobility in two different gels can be described by

$$\ell og\ \frac{m_1}{m_2} = (k_c MW + A)(T_2 - T_1) \qquad (7)$$

$$= \alpha MW + \beta$$

This relationship is useful for determining the molecular weight of proteins that have to be analyzed under nondenaturing conditions (e.g., enzymes).

In SDS electrophoresis, differences in charge and

shape (hence, $m_0$) of the proteins to be resolved are minimized by reducing with a thiol reagent and complexing with the detergent anion, dodecyl sulfate; the detergent binds to proteins at an approximate ratio of 0.5 molecules/amino acid residue, unraveling them to an extended, rodlike form and overwhelming their charge with its own negative charge [23]. It is expected, therefore, that, for protein-SDS complexes, $m_0$ will be uniform and, therefore, from Eqs. (5) and (6), or Eq. (7),

$$\ell og\ m = \gamma MW + \delta \qquad (8)$$

where $\gamma$ and $\delta$ are gel-dependent constants. Surprisingly, all reports in the literature, beginning with the earliest work of Shapiro et al. [24], indicate a different relationship,

$$\ell og\ MW = \gamma' m + \delta' \qquad (9)$$

The paradox may be explained by the fact that, irrespective of which of the two equations is correct, a given set of values of (m,MW) can also give a satisfactory (within experimental error) fit to the other equation, provided the range of values is not too wide. A similar uncertainty holds in the case of non-SDS gels [22], for which an alternative to Eq. (7) has also been offered [25].

$$\ell og\ MW = \alpha' \frac{m_2}{m_1} + \beta' \qquad (10)$$

In practice, it is found that in SDS-polyacrylamide gels mobility can be linearly related to $\ell og$ MW, within a relatively narrow molecular weight range, which can be shifted downward by increasing T or C. At the extremes of the range, curvature appears and is particularly severe at the low molecular weight end (Fig. 6). This would be predicted if, in fact, Eq. (8), rather than Eq. (9) is correct (although alternative explanations exist). At the extremes of the useful range, it may be helpful to apply both equations. In any case, the best solution is to adjust the gel parameters, T and C, so as to bring the proteins to be resolved within the useful range of the gel. If the proteins to be resolved fall in an unusually wide molecular weight range, gels with a gradient of T and/or C should be used, and sufficiently numerous standards should be employed to define the standard curve.

Abnormally low mobility is shown by proteins that bind reduced amounts of SDS (e.g., glycoproteins, chemically cross-linked proteins). Although the intrinsic charge of the protein will usually be less than 10% of the charge imparted by the SDS, it will be more significant in the case of highly charged proteins and may lead to deviations from linearity (e.g., histones show atypically low mobility; see Ref. 23). As expected, deviations are particularly noticeable with small proteins, in which absolutely small changes in intrinsic charge or size are relatively more significant. Gels with high C values, which contain 8M urea, should be used for small proteins, as recommended by Swank and Munkres [26].

Resolution. A critical factor affecting resolution in gel electrophoresis is the narrowness of the protein bands as they enter the gel. Very narrow bands can be attained with stacking gels [27,28]. The stacking modification of SDS gels [29] has led to significant improvement in resolution [30]. It should be recalled, however, that even without special stacking gels narrow starting bands result (a) whenever the sample is applied in a very small volume, (b) whenever the proteins are greatly retarded as they enter the gel, that is, whenever $m/m_0$ is very small (hence the desirability of using the highest T and C values consistent with linear and convenient resolution of the proteins under study), and (c) whenever the proteins are forced to carry a large fraction of the current in the applied aqueous sample, by reducing the ionic strength of the sample buffer to a minimum i.e., below that of the gel and running buffers (in that case low current should be used at first, until the proteins enter the gel, in order to avoid overheating).

In the low porosity gels in common use, diffusion is usually not a major problem. The commonest sources of reduced resolution are distortion of the bands due to overloading and curvature due to overheating.

In regions of high protein concentration, the buffer concentration and hence the conductivity are correspondingly lower. Therefore, the electric field strength increases locally and the protein molecules move faster. The protein distribution is thus progressively converted from a Gaussian to an asymmetric form, as molecules originally in the middle of the distribution "overtake" the leading edge (Fig. 7).

The parameter that favors this distortion is the ratio of the macromolecule mass divided by the ionic strength. Also, overloading becomes more serious as the charge of the macromolecule increases and its mobility decreases (i.e., as the macromolecule is more strongly affected by E without itself contributing to the conductivity); hence the danger of overloading is particularly serious for RNA and for proteins in highly retarding, high T and/or C SDS gels. When the loading is high, as in preparative gels, the ionic strength should be correspondingly increased; to prevent overheating, the voltage should then be lowered and electrophoresis performed at low current for a long period (even several days).

If the power passing through a cylindrical gel is too high, overheating occurs. Because of the geometry, heat is least effectively dissipated from the central axis of the gel, and most effectively from the periphery.

As heating increases, the viscosity and hence the frictional resistance, f, decreases; thus, mobility increases. As a result, the band becomes deformed, with its center curved forward in the direction of migration.

<u>Procedures for SDS Electrophoresis of Silk Moth Chorion Proteins (M. Paul and J.C. Regier; see also Ref. 5).</u> For most proteins, the analytic method of choice for fractionating by molecular weight is the stacking SDS slab gel procedure [30]. Because of their unusually small size and low solubility chorion proteins show an improved resolution in gels made according to a slight modification of the Swank and Munkres [26] procedure. These are highly cross-linked gels (C = 7.1%) which also contain 6M urea. To avoid artifacts and loss of resolution by reoxidation of the proteins, the gel is cleared of oxidizing agent by electrophoresis in the presence of mercaptoacetic acid. Alternatively (or in addition), sulfhydryl groups are blocked by carbamidomethylation, as previously described. Other general precautions to be followed in sample preparation are given by Weber et al. [23].

In general, slabs are preferable to cylindrical gels, because of the convenience of comparing samples and of evaluating radioactiviy profiles by autoradiography. Unfortunately, high concentration gels, and especially those that are highly cross-linked, are very difficult to dry. Moreover, with chorion proteins we have not been able to attain as high a resolution on slabs as on cylinders. Therefore, we routinely use cylindrical gels, as follows:

Gel Preparation. Final concentrations are 13% acrylamide, 1% bis, 6M urea, and 0.5% SDS. Purity of the components are as given for isoelectric focusing. The solution should be made up just before use. To make 30 ml (150 cm of gel column, i.e., fifteen 10-cm gels or the equivalent), mix:

3.90 g acrylamide
0.30 g bis
0.15 g SDS
10.8 g urea
11.9 ml $H_2O$
6.0 ml Tris-Acetate buffer. This is a stock solution containing 0.5M HAc adjusted to pH 8.4 with Tris base (0.716 ml glacial HAc/25 ml).

This solution is filtered through Whatman #1 filter paper and degassed briefly, and 10 mg ammonium persulfate in 25 μl $H_2O$ and 2 μl TEMED catalyst are added. The gels are made in siliconized (see Sec. V.B) 5 mm I.D. lime glass tubes by standard procedures. Polymerization at room temperature usually takes about thirty minutes. For conveniently obtaining a sharp meniscus, it is possible to overlayer with isobutanol instead of water.

Reservoir Buffer. This is 0.1M HAc and 0.1% SDS, adjusted to pH 8.4 with Tris base (5.72 ml glacial HAc/liter), and supplemented with 3.73 ml 95% mercapto-acetic acid (final concentration 0.053M and pH 8.2).

Sample Preparation. The sample is prepared as for isoelectric focusing, and carbamidomethylated. Finally, SDS is added to 1%. It is possible to decrease the amount of DTT further, and the amount of iodoacetamide correspondingly. This is particularly useful if [$^{14}C$] iodoacetamide is to be used for in vitro labeling. For dissolving Bombyx chorions at least 0.03M DTT is needed and, for polyphemus chorion, 0.01M.

We load 10-20 μg chorion protein per gel. One mature polyphemus chorion has about 450 μg protein; dissolved in 100 μl buffer, the mass ratio SDS:protein is 2.2:1. Earlier chorions have less protein. For developmental series of polyphemus chorions, the fraction of chorion per gel can be adjusted according to position within the ovariole [5] as follows: position three, 0.5; four, 0.33; five, 0.20; six, 0.10; eight and up, 0.05. As much as one half of a follicular epithelium can be added per gel. These values can be increased by a factor of ten for Bombyx samples.

Gel Running. Samples are added to the top of the gel, under reservoir buffer, with a Yankee Disposable Micropipet; delivery is controlled with a Clay Adams Safety Pipet-Filler (VWR Scientific). The samples are dense, so they run down and spread out evenly over the gel top. A reducing environment is produced during the run by the mercaptoacetic acid front. As little as 5 μl can be added by this technique reasonably quantitatively. Bromophenol blue tracking dye (K & K Laboratories, Inc.) may be added to the sample before loading or to a blank gel. It is essential to use constant and very low current: we run 10-cm gels for approximately 56 hr at 1 mA/gel. The dye band should move at a rate of about 0.5 cm/hr. In these gels, the lowest molecular weight chorion component in polyphemus ($A_1$, MW about 7,000) migrates with a mobility of 0.35-0.40 relative to bromophenol blue. Determine the time required for tracker dye to migrate to the desired position of chorion component $A_1$, then multiply this time by 2.7 to determine the duration of the run. Remove gels from the tubes by air pressure, then fix and stain by standard procedures (see isoelectric focusing).

For radioactivity counting, we slice the gels with a homemade slicer made of razor blades held apart with 1 mm washers; the entire assembly is held within a brass cover. The gel is positioned on a rubber trough (made of rubber tubing cut in half longitudinally) and the slicer is pressed down slowly but firmly. For lower T or C gels, it may be necessary to freeze the gel first. Individual slices are put in 5-ml disposable vials, 4 ml of fluor is added (42 ml Liquifluor, 30 ml Protosol in 1 liter toluene), the vials are capped (use aluminum foil-lined caps, Arthur Thomas Co.), warmed at 45 $^{o}$ C for 10 to 12 hrs, chilled, vortexed and counted by liquid scintillation.

Cylindrical Gradient Gels (A. Wyman and W. Petri, unpublished). When fractionating mixtures of proteins ranging widely in molecular weight, gradient rather than uniform gels may be used. For Drosophila chorion proteins, we have used T = 7.5 to 14% or 5.4 to 14% gels (C = 7.1%). The procedures for the 5.4 to 14% gels are as follows:

Gel Tube Assembly. Twelve gel tubes (12 cm x 7 mm O.D.) are placed in an open glass cylinder (14 cm x 29 mm I.D.). The top of the cylinder is partially covered with a bar of Lucite and the bottom is sealed with a 4.5-cm rubber square supported on a piece of Lucite. These pieces are held tightly in place with a large "C"-clamp fastened to a ring stand. About 5 ml of isobutanol is delivered into the cylinder.

Acrylamide Stock. 50 g acrylamide and 3.85 g bis, brought to 100 ml with distilled $H_2O$.

Gel Buffer. 3.5 g Tris base, brought to pH 8.4 with glacial HAc and to 50 ml with distilled $H_2O$.

Reservoir Buffer. 2 g SDS and 54 g Tris base, adjusted to pH 8.4 with glacial HAc (about 4 ml) and brought to 2000 ml with distilled $H_2O$. For upper bath, add 2.1 ml of 80% mercaptoacetic acid/liter.

Solutions for Making Gels:

| Ingredient | 5.4% Solution | 14% Solution |
|---|---|---|
| Urea | 7.2 g | 7.2 g |
| Acrylamide stock | 2.0 ml | 5.2 ml |
| Tris-acetate buffer | 4.0 ml | 4.0 ml |
| SDS (100 mg/ml) | 1.0 ml | 1.0 ml |
| Sucrose (80%) | 2.5 ml | 5.0 ml |
| Water | 5.0 ml | -- |
| TEMED (10%) | 30 μl | 25 μl |
| Ammonium persulfate (5%)* | 25 μl | 20 μl |

* added last

To make: Mix urea, acrylamide, buffer, SDS, sucrose and water (final volume, 20.0 ml each solution). Filter, degas. Add TEMED and ammonium persulfate. Mix. With 5.4% solution in the mixing chamber of a small gradient maker, allow acrylamide mixture to flow through tubing inserted between the gel tubes into the bottom of the cylinder. Start slowly. The isobutanol will rise, remaining on top of the gel solution. When polymerization is complete (1-2 hours), remove cylinder from clamp, push out gels, and rinse off. Gels can be stored for several days at room temperature, if they are water-tight. For running *Drosophila* chorion, use the formula mA-hr/gel = 16, with 8 to 16 hours as the time. Congo red can also be used as a second visible marker, in addition to bromphenol blue; it migrates as a small protein, the exact mobility depending on the gel system.

## III. DETERMINATION OF RATES OF SYNTHESIS OF SPECIFIC PROTEINS

### A. General

Central to our approach is the evaluation of rates of protein synthesis. Different methodologies are applicable, depending on how the information obtained is to be used. In general, estimates of relative rates of protein synthesis are straightforward; estimates of absolute rates are not. We will first discuss two approaches to the study of relative synthetic rates. In the first, the synthesis of particular components is evaluated in each of a series of samples, relative to other components in the same sample; in the second, the relative synthetic rates of a large number of components in one sample are compared to the corresponding rates in a second sample. Single-isotope and double-isotope procedures are best adapted to these two approaches, respectively. Thus, the second approach usually requires liquid scintillation counting for the determination of two isotopes, whereas the first is compatible with the convenient autoradiographic detection of one isotope. We will finally discuss determination of absolute rates by our procedures.

### B. Internally Calibrated Relative Rates of Protein Synthesis

These are the easiest rates to evaluate. In essence,

one asks, with a series of samples, how prominent a labeled protein is relative to other proteins labeled in the same sample. This permits one, for example, to estimate, as a function of developmental time, the rate of synthesis of a particular protein (or set of proteins) relative to the synthesis of other protein "markers" or of total protein. One merely has to briefly expose the cells to labeling medium containing one radioactive amino acid, extract and fractionate the proteins, and then evaluate the incorporation into each desired fraction. For example, by using a series of progressively older follicular epithelia, pulse-labeling each one with the same amino acid, and fractionating its labeled proteins on SDS-polyacrylamide gels designed to separate chorion from non-chorion proteins (which are larger), we determined that the rate of chorion protein synthesis increases from undetectable to over 90% of the total in approximately 9 hours [5,6]. For greatest accuracy, it would be desirable to take into account the amino acid composition of the proteins being studied, although this will normally introduce only a proportionality factor.

In this approach, the kinetics of labeling of the intracellular precursor pool are not important, provided that one assumes that the synthesis of all the proteins being studied draws on the same intracellular amino acid pool. This is a reasonable assumption, to the extent that one compares proteins synthesized by the same cells. Even homogenous populations may change their pool characteristics with developmental time or experimental conditions. Thus, it is advisable to use single amino acids rather than mixtures for labeling. If amino acid mixtures are used, a series of samples may show differences in the relative rates of incorporation into two proteins (dissimilar in amino acid composition), merely because of nonproportional labeling of the pools, rather than because of true changes in the relative rates of synthesis. If it is essential to use heterogeneous cell populations, one should avoid comparing the synthesis of proteins produced by different cells, since the subpopulations may also differ in terms of pool characteristics (kinetics of uptake and the specific activity of the intracellular label); failing that, there may be some advantage in labeling with amino acid mixtures, which may dampen variations in pool specific activities and protein amino acid compositions. In this case, however, the results should frankly be recognized as only semiquantitative.

In experiments of this type, it is necessary to take into account the possibility of variable protein turnover or post-translational modification. For example, with a long pulse, part of the labeled protein in question may disappear from its expected position in the fractionated protein profile, because of unusually rapid turnover or secondary modification. This problem can be elucidated by comparing identical samples pulsed for widely different durations or exposed to different pulse-chase regimens. Rapid turnover, or secondary modification (to a form resolvable by the fractionation procedure used), will lead to the progressive disappearance of the labeled protein, and its relative rate of synthesis will be underestimated.

If one wants to study the relative synthetic rates of a prominent component, the most convenient procedure is to dissolve all proteins by treating the cells with an SDS- mercaptoethanolmixture, fractionate on an SDS-polyacrylamide slab, perform autoradiography (or fluorography), and quantitate by densitometry. In this way, quantitation is possible against a whole set of internal "markers," all analyzed as part of a single sample. If autoradiography is not convenient (e.g., because of the difficulty of drying low-porosity gels, such as those needed for chorion resolution), the protein mixture can be analyzed in a cylindrical gel and the radioactivity determined by slicing and counting by liquid scintillation. We used this method for defining the developmental stages of choriogenesis, in terms of "protein synthetic profiles" [6]. Autoradiography or fluorography can always be used if the proteins are fractionated on IF slab gels. Of course, one can only standardize against the synthesis of proteins that can be resolved in the same ampholine range; moreover, it is necessary to seriously consider the possibility of post-translational modification.

For studying relative synthetic rates of a minor component, a single fractionation step may be inadequate for accurate quantitation against a background "smear." One may then resort to a two-dimensional gel system [14]. Alternatively, or in addition, a pre-purification step may be interposed, such as precipitation with a specific antibody, prior to the final electrophoretic display. In this case, one should monitor losses, by incorporating into the tissue extract a known amount of the same protein labeled with a different isotope ("standard"); this leads to a double-isotope experiment (see below). For an application of this procedure to the study of relative synthetic rates over development, and for conditions optimizing specific immunoprecipitation, see Refs. 31 and 32.

### C. Detailed Comparison of Two Synthetic Profiles: Double-Labeling Analysis

For some applications, it is desirable to compare quantitatively the relative rates of synthesis of a number of proteins in two different samples. Although this can be achieved by parallel analysis of the labeled samples in a single slab gel, greater confidence is possible if the two samples are mixed prior to fractionation.

This avoids the possibility of artifacts due to differential losses or electrophoretic irreproducibility. The samples can be mixed if they are labeled with two different isotopes, since each isotope can be detected separately by liquid scintillation. Examples of this approach are the comparison of synthetic profiles of a single follicle at the beginning and end of an in vitro culture experiment [6] or comparison of synthetic profiles before and after treatment of cells with actinomycin D [32,33] . In general, double-labeling is a very effective approach, ideal for picking up small differences between otherwise matched samples (e.g., see [34]). It also permits use of the very same cells at two time points, thus decreasing noise due to a biologic variability. For example, synthetic maturation in culture was documented and timed with individual follicles by this method [6].

In this method, changes in relative rates of synthesis are immediately apparent from the fact that the fraction in question has an unusual isotope ratio. Needless to say, the samples should be counted by liquid scintillation under identical conditions (constant and known quenching, preferably estimated by the external standard method). Constant-ratio controls should always be included (two identical samples, labeled with different isotopes), and the isotopes should be switched between experimental samples for added confidence. Deviations of up to approximately 10% from ratio constancy can be expected in control samples. Sufficient counts should be collected to prevent added uncertainty due to counting errors. We have designed a computer program for processing and plotting double-labeling data, including ratios [35]. Ratios are not well suited to statistical evaluation, but are convenient for interpretation; an alternative double-label program has been described [36], using regression analysis.

The choice of isotopes is essentially limited to $^{3}H$ and $^{14}C$, in the case of proteins, although $^{32}P$ and $^{33}P$ can also be used in the case of nucleic acids. If the amino acid can be cysteine or methionine, $^{35}S$ can be substituted for $^{14}C$. As discussed previously, it is essential to use the same amino acid for both samples, and single rather than mixed amino acids (to avoid pool complications). If that is impossible, controls with the isotopes switched are mandatory. The choice of an average isotope ratio should be determined by the need to minimize counting error and error due to crossover (quench) variation. We recommend $^{3}H/^{14}C$ ratios of at least 1 and up to 10, although there is essentially no upper limit with a properly adjusted $^{14}C$ window.

### D. Determination of Absolute Rates

This depends on knowledge of the specific activity (SA) of the intracellular labeled precursor pool (SA, counts per minute per mole). Once the SA is known, absolute rates can be evaluated:

Synthetic rate = incorporation rate/SA

Here SA can be surmised from kinetic analysis of incorporation data [37], or from chemical determination on the extractable precursor. A number of methods are possible for the chemical determination of SA. Basically, one wants a measurement of counts per minute (cpm) associated with a known number of precursor molecules. For amino acids, we prefer our double-label micromethod [38], in which the number of amino acid molecules are estimated by reaction with [$^{14}$C]dinitrofluorobenzene. For example, in an experiment involving incorporation of [$^{3}$H]leucine, tissue SA is estimated by reacting the total amino acids with [$^{14}$C]DNFB, purifying the DNP-leucine by two-dimensional thin layer chromatography (TLC), and determining its radioactivity by two-isotope liquid scintillation procedures. Thus, in a single sample we determine both the number of precursor molecules ($^{14}$C) and the isotope they carry ($^{3}$H). The SA is determined directly, and is unaffected by losses or incompleteness of the reaction; for absolute values, one only needs to know counting efficiencies and the SA of the DNFB (which can itself be determined). The original publication should be consulted for technical details and for further discussion. It should be pointed out that the procedure is exceedingly sensitive, can be used for SA values ranging over approximately six orders of magnitude, is very easy to run once it is set up, and can be modified for different amino acids (we use leucine) and for different derivatives (e.g., with dansyl chloride; see Ref. 39). We should caution that, although DNP-leucine and DNP-isoleucine can be separated by continuous chromatography in one dimension [40], a smear of [$^{14}$C]DNFB derivatives exists in the area of these spots and can lead to significant errors if undetected; we routinely use two-dimensional TLC separations for greater accuracy.

The drawback of chemical determination of SA is that one must assume that the total extractable pool is the immediate precursor used in protein synthesis. This need not be the case, since amino acid compartments may exist within the cell. On the other hand, the kinetic evaluation of SA is indirect and depends on certain assumptions. The solution is to extend the chemical methodology to permit SA determination of what are known to be immediate precursor pools: the nascent polypeptides or peptidyl-tRNA and (less stringently) the aminoacyl-tRNA. The latter has recently been accomplished [41].

## IV. IN VITRO LABELING OF PROTEINS

### A. General

The in vitro introduction of covalently bound radioactive label into protein molecules may be used for various purposes. These include functional analysis (e.g., probing the active center of enzymes), direct quantification of proteins present in minute amounts (e.g., by stoichiometric labeling of the active serine or proteases by [$^3$H]DFP), indirect use in quantitation studies (e.g., isotope dilution or radioimmunoassay utilizing iodinated proteins). Perhaps our commonest purpose is the convenient preparation of a labeled standard, to be used for comparison with fractionated proteins of an experimental sample or to follow a purification procedure. For comparison with experimental samples, the in vitro labeled standard can be run in parallel slots of a slab gel, or it can be mixed with the unknown, if different isotopes have been used (internal standard); prior to analysis, the experimental proteins must be modified with the same reagent as that used in in vitro labeling (non-radioactive, of course), unless the modification introduced by in vitro labeling can be shown to be inconsequential for the fractionation scheme used (this is usually the case for SDS gel electrophoresis, but not for isoelectric focusing).

General rules to be remembered are that few reagents are completely specific; that some reactable groups may require denaturation to be exposed; and that if native proteins are labeled conformational changes may ensue, especially with reagents that significantly affect the pK of the groups modified.

Iodination is a widely used method [42], which will not be discussed here. Maleylation [43] is useful for purification work, because it is reversible at acid pH and introduces charge changes; stable acylation is also possible, with such reagents as succinic anhydride and acetic anhydride. Dansylation [44,45] is a quick and simple procedure, which is particularly attractive because of the ease of detecting the derivatives; reaction with fluorescamine [46] has similar advantages. Other than iodination, we have made the greatest use of reductive methylation and of alkylation with iodoacetamide (carbamidomethylation). A general review concerning chemical modifications of proteins has been published by Means and Feeney [47].

## B. Reductive Methylation

Formaldehyde is available in labeled form, both with $^{3}H$ and with $^{14}C$. Depending on the lysine content, proteins can be labeled with these reagents to the extent of about 6000 cpm/μg and 2000 cpm/mg, respectively. For specific activities two orders of magnitude higher, [$^{3}H$] $NaBH_4$ can be used. Formaldehyde reacts with amino groups [48-50], adding $>CH_2$, which is reduced by borohydride to a methyl group. The reaction is primarily with unprotonated amines, and dimethylation is possible, depending on the conditions. Reactive groups are primarily the $\varepsilon$-$NH_2$ of lysine and the $\alpha$-$NH_2$ of the N-terminus, under sufficiently basic conditions. A large excess of formaldehyde should be avoided, to minimize reaction with histidine, and the borohydride should not be in molar excess, to decrease the likelihood of disulfide reduction and peptide bond cleavage. For optimal results, use molar ratios of $CH_2O$/Lys residue = 7 to 9 and $CH_2O/NaBH_4$ = 2. The experimental protocol is as follows:

1. Keep buffers and reagents and perform all operations at $0^\circ$ C
2. Add protein in a small volume of buffer (0.1 to 1 ml, 0.2M sodium borate, pH 9.5, or 0.2M sodium carbonate, pH 10)
3. Add formaldehyde as 1 to 4M solution (usually a few microliters). Use the radioactive formaldehyde as supplied. Do not dilute.
4. Wait 30 sec and make 10 additions of a solution of $NaBH_4$ (a few microliters each time) in a period of 30 min (one addition every 3 min); $NaBH_4$ is dissolved in the high pH buffer as 0.1-0.4M solution.
5. Just 3 min after the last addition, add 1N HCl to adjust the pH to 5.5, to destroy the excess borohydride.
6. Desalt by gel filtration on a 1 x 25 cm column of Sephadex G-25 or G-50, equilibrated and eluted with 1mM phosphate buffer (pH 7.5) at $4^\circ$ C. Omission of the pH adjustment to 5.5 before the gel filtration permits reversibly bound formaldehyde to be retained on the protein.

### C. Carboxymethylation and Carbamidomethylation

Alkylation with iodoacetic acid has the disadvantage of introducing negative charges and the advantage of availability of the labeled reagent at relatively high specific activity (both $^3H$ and $^{14}C$). We use iodoacetamide ($ICH_2CONH_2$), which is available in the $^{14}C$-labeled form, unfortunately only at 20mCi/mmole. For proteins containing 2% Cys, in vitro labeling can give specific activities as high as 8000 cpm/μg. Iodoacetamide [51] can react with histidine (at pH about 5.5), methionine (at pH 2-8.5, but slowly), and amino groups (at pH greater than 8.5). At pH 3, it is possible to label methionine specifically, albeit slowly. Under the conditions described below, iodoacetamide reacts only with cysteine (provided the reaction does not continue beyond 15 min). The reaction is quantitative, because of the denaturing conditions. Since the pK of the sulfhydryl is high, under acid conditions the modified proteins have the same charge as they do in their native form. Thus, iodoacetamide is an excellent label for silk moth chorion proteins, since it introduces substantial radioactivity (because of the high cysteine content) without affecting the isoelectric point. The protocol we use is as follows:

1. Make 10 ml 6M guanidine-HCl, 0.36M Tris (pH 8.4), 0.75mM $Na_2$ EDTA, 0.03M DTT.
2. Add to 10-100 mg protein under a nitrogen barrier. Keep at room temperature for several hours.
3. Dissolve 64 mg iodoacetamide (0.345 mM) in 1 ml 0.36M Tris (pH 8.4). Add this to the protein solution. This and subsequent steps should be performed in the dark (because of the $I^-$ generated).
4. React at room temperature for 15 min.
5. Add 0.1 ml mercaptoethanol. Adjust pH to <5.
6. Desalt with column or by dialysis.

## V. PURIFICATION OF CHORION mRNA

### A. Dissection of Animals

Commercially raised pupae of *Antheraea pernyi* or *Antheraea polyphemus* are stored at 2° C until needed. Development to the adult moth is begun after transferring to 25° C. Females are recognizable by their narrow antennae and characteristic genitalia. Developmental stages are determined as described [3]. Follicles are removed on day 15 or 16 of adult development, when the face is dark pink and the animal has become soft through partial digestion of the pupal cuticle.

Animals can be dissected directly or after brief carbon dioxide anesthesia. The abdomen is cut open from the back, using scissors and precautions not to pierce the midgut and hindgut must be taken. The eight follicle-containing ovarioles are removed carefully with fine forceps and placed into a Petri dish containing sterile Grace's medium [52], with crystals of phenylthiourea to prevent melanization. Chorionating follicles are identified by their position within the ovariole [5]. Follicles from positions 2 to 10 (or up to ovulation) are separated from the rest under a dissecting microscope, transferred to another dish, and freed of tracheae and the ovariole sheath. The follicles can be used directly or stored in liquid nitrogen until needed. We usually start a message preparation (which will eventually lead to about 30 μg mRNA) with 1000 follicles (about 14 animals).

### B. Magnesium Precipitation and Extraction of RNA

As a first step toward isolation of mRNA, a fraction rich in polysomes is prepared by the $Mg^{2+}$ precipitation technique [53]. All solutions, and containers, must be sterile. We use almost exclusively 30 ml siliconized Corex tubes.

Siliconization (in a hood) is performed as follows: clean and dried tubes are filled with a (reusable) solution of 5% (v/v) dichlorodimethylsilane, in chloroform. Because of toxicity, inhalation of the chemical and contact with the skin should be avoided. After a minute, the tubes are drained, left in the hood for a few minutes to dry completely, and then washed thoroughly with distilled water, dried, and autoclaved.

Buffers A, B, and C are prepared from sterilized stock solutions, as shown in Table 2. Triton, heparin, and sucrose are not sterilized, but are added directly or dissolved in sterile water (sucrose).

All operations are performed in the cold room.

One thousand follicles are ground with a mortar and pestle in 40 ml buffer A. Because of the tough chorion, a Dounce homogenizer cannot be used.

TABLE 2

Buffers Used in the $Mg^{2+}$ Precipitation Procedure

| | Stock (M) | To Add (ml) | | | Final Concentration (mM) | | |
|---|---|---|---|---|---|---|---|
| Buffer | | A | B | C | A | B | C |
| NaCl | 1 | 2.5 | 2.5 | 2.5 | 25 | 25 | 25 |
| $MgCl_2$ | 1 | 0.5 | 20.0 | 0.5 | 5 | 200 | 5 |
| Tris-HCl, pH 7.5 | 0.5 | 5.0 | 5.0 | 5.0 | 25 | 25 | 25 |
| Triton X-100 | -- | 2.0 | 2.0 | -- | 2[b] | 2[b] | -- |
| Heparin | -- | 100[a] | 100[a] | -- | 1[c] | 1[c] | -- |
| Sucrose | 2.0 | -- | -- | 50.0 | -- | -- | 1000 |
| Water | -- | 90.0 | 70.5 | 42.0 | -- | -- | -- |

[a]In milligrams
[b]In percent
[c]In milligrams per milliliter

The homogenate is transferred to cold tubes and centrifuged for 5 minutes at 27,000$g_{max}$ (15,000 rpm for the IEC B-20 centrifuge) at 0° C. The supernatant (about 50 ml) is decanted into a beaker, an equal volume of buffer is added, and the mixture is kept on ice for at least one hour. Subsequently, the mixture is layered very slowly over one-third volume buffer C (six tubes are used) and spun for 10 minutes at 27,000$g_{max}$. The tubes are inverted abruptly to decant the supernatant and any lipid or sucrose left on the walls is wiped with tissue paper. Small amounts of buffer C remain at the bottom of the tube, because the polysome pellet is very loose and cannot be washed.

Each of the six pellets is suspended by vortexing in 2 ml 0.02M Hepes (adjusted with KOH to pH 7.5). An equal volume of 0.2M NaAc (pH 5.0) is added and the tubes are transferred to room temperature. The SDS is added immediately to 0.5% and the tubes are warmed to room temperature and shaken. An equal volume (4 ml) of Sevag (chloroform-isoamyl alcohol 24:1, v/v) is added, and each tube is vortexed for a few minutes. The mixture is centrifuged briefly in the bench centrifuge to separate the phases, and the upper aqueous phase is transferred to clean tubes. The organic phase is extracted with 2 ml Hepes-NaAc, and, after centrifugation, the aqueous phases are combined. An equal volume of phenol is then added and, after vortexing, a volume of Sevag. After vortexing and centrifugation, we discard the organic phase and repeat the last step. The aqueous phase is transferred to clean tubes and extracted once with Sevag and twice with ether. Any ether residue is evaporated under nitrogen, and the RNA is precipitated with two volumes of 95% ethanol at $-20^{\circ}$ C.

Some aspects of this procedure need further discussion.

It is possible to cut the follicles in half to remove the yolky oocyte and its RNA. But to deyolk 1000 follicles is a tedious, time-consuming, and unnecessary procedure. The yolk does not interfere with the $Mg^{2+}$ precipitation, and the absence of serious contamination of the final chorion mRNAs with other messages is suggested by the coincidence of stained RNA bands and autoradiographically detectable pulse-labeled chorion mRNAs from follicular cells. In addition, immunoprecipitable chorion polypeptides account for much of the cell-free translation product obtained with our final mRNA preparation.

At a concentration of 0.1M, $Mg^{2+}$ precipitates ribonucleoprotein complexes and even naked high molecular weight RNA (but not tRNA). It also inhibits ribonuclease. Although the follicular epithelium is a tissue with very low ribonuclease activity, heparin, a ribonuclease inhibitor, is included in buffers as an additional precaution. In our hands, RNA extraction at pH 5 gives better yields than extraction at pH 9 [54].

Use of phenol without Sevag should be avoided because phenol alone does not efficiently release poly(A)-containing RNA into the aqueous phase [55]. But, we observed that vortexing with phenol alone followed by the addition of Sevag is more efficient than using a phenol-Sevag mixture directly. The preliminary treatment with Sevag alone is important because it removes most of the SDS. Without this treatment, the protein-SDS complexes produce a huge interphase, and the aqueous phase decreases to less than one third of its initial volume; efficient extraction of RNA then requires repeated extraction. Phenol contaminating the precipitated RNA is effectively removed with ether.

For quantitative ethanol precipitation of RNA, 2-6 hr in the cold are generally recommended. Almost instantaneous precipitation can be achieved, however, by cooling the RNA solution (in 0.5M $NH_4Ac$ or 0.1-0.25M NaAc) in a dry ice-acetone bath (A. Maxam, personal communication). Using this technique with [$^{32}$P]RNA, we observed quantitative precipitation even from very dilute solutions, which yielded no visible pellet.

### C. Oligo(dT)-Cellulose Chromatography

The occurrence of poly(A) in mRNA molecules can be exploited for their purification from ribosomal and other RNAs, by a variety of procedures:

Oligo(dT)-cellulose chromatography [56, 57]
Poly(U)-Sepharose chromatography [58]
Chromatography on unmodified cellulose
Binding to cellulose nitrate membrane filters (Millipore filters) [64] or poly(U)-glass fiber filters [65]
Hydroxyapatite chromatography after hybridization in solution with poly(U) [66]

Our experience is with oligo(dT)-cellulose chromatography, which seems the method of choice for large-scale preparation of poly(A)-containing RNA. Poly(U)-Sepharose columns work equally well, but a small amount of poly(U) is eluted with the RNA. Use of plain cellulose for mRNA purification has serious disadvantages: the columns are not reusable and the highest binding capacity observed with commercially available celluloses was only about 50 μg poly(A)/g cellulose (probably a function of lignin content). It should be noted, also, that single stranded DNA binds to cellulose ( like poly(A)). Filters (especially poly(U) glass fiber filters) can be used as a rapid assay method for poly(A)containing molecules in an RNA population. Millipore filters are claimed not to be very effective when the poly(A) sequences are short ([54]; but see Ref. 10; they also bind other polynucleotides (e.g., poly(U) and poly(G)). Finally, hydroxyapatite chromatography has the drawback that the poly(A)-containing RNA is contaminated with poly(U) so it can be recovered only after dialysis or precipitation with CTAB (since it is eluted with phosphate buffer, which precludes ethanol precipitation).

Though oligo(dT)-cellulose can be synthesized relatively easily [67, 68], commercial preparations are convenient. The binding capacity differs tremendously from company to company, and batch to batch; new batches should be tested before use. The best oligo(dT)-celluloses tested so far in our laboratory were from Amersham (batch 3F) and Collaborative Research (Type T-3, Lot 534-35); they bind about 100 μg poly(A)/100 mg resin. Assuming that mRNA is 1-2% of polysomal RNA, and poly(A) 10-20% of mRNA, 100 mg of this oligo(dT)-cellulose should bind all the mRNA from 25-100 mg of polysomal RNA. One of the greatest advantages of this method is that the columns are reusable for a very long time. A sample can also be processed batchwise [69-71], permitting elution of minute amounts of message in very small volumes.

We use the following procedure, at room temperature and under sterile conditions: approximately 200 mg oligo-(dT)-cellulose are suspended in binding buffer (0.5M KCl, 0.01M Tris, pH 7.5). If the presence of SDS is desired as an extra precaution against ribonuclease, the buffer can be 0.5M NaCl, 0.01M Tris, 0.2% SDS. We observed no difference in the effectiveness of the two buffers (in contrast with the results reported for plain cellulose; [61]).

The suspension of oligo(dT)-cellulose is packed, under slight pressure, into a small polypropylene column (0.7 x 4 cm, with a reservoir at the top; Bio-Rad), forming a 0.7 x 1.5 cm bed. Packing and operating the column under pressure has the advantage of an adjustable flow.

The column is washed with binding buffer until $A_{260}$ = 0 and then calibrated. The binding capacity is measured by passing through a known amount of poly(A) and determining the fraction that binds. It is convenient to use radioactive poly(A), especially the $^{32}P$-labeled form, which can be evaluated by Cerenkov counting and reused. The flow rate should be very slow (about 0.25 ml/min). The performance of the column seems to be concentration independent, at least for poly(A) or chorion polysomal RNA. For this and other procedures, we have used rabbit globin mRNA as a convenient test material. With globin polysomal RNA, which easily aggregates, we have observed clogging of the column at high concentrations. Therefore, a concentration not higher than 2-3 mg/ml is generally recommended.

The column is washed with binding buffer until no radioactivity or UV-absorbing material is eluted (10 ml is usually sufficient). The bound material is eluted with $H_2O$ (or 0.01M Tris, pH 7.5), usually in 4 to 6 ml. The column can be regenerated for repeated use with 0.1N KOH (or NaOH). It should be stored in the cold in 0.1N KOH and washed extensively with binding buffer before use.

There is no general agreement over the temperature of this operation. It is possible to run the binding step in the cold and then elute the bound material at room temperature or at 37° C, using a water-jacketed column. We find room temperature operation entirely satisfactory.

Aviv and Leder [56] described a two-step elution of bound material, first with 0.1M KCl, 0.01M Tris, and then with Tris alone. We do not recommend this procedure, since in our experience the material eluting at intermediate ionic strength contains substantial amounts of chorion mRNA -- possibly the molecules with particularly short poly(A) sequences [10].

Some material (about one-tenth of the total bound) sticks to the column tenaciously and can be eluted only with base (naturally, it is discarded).

Although oligo(dT)-cellulose chromatography is an excellent technique for the enrichment of an RNA sample in mRNA, it should be stressed that it does not purify mRNA completely. Gielen et al.[57] reported that, in the bound fraction isolated from rabbit reticulocyte polysomal RNA, 40-70% of the RNA was heavier than 9S. With both rabbit globin and chorion mRNAs, we find that 75-80% of the bound fraction consists of contaminants, after a single passage through the column. We find it necessary to use two binding steps, separated by one size fractionation (via sucrose gradient). The bound material recovered from the first column is precipitated by ethanol dissolved in "SDS buffer" (0.5% SDS, 0.1M NaCl, 0.01M Tris-HCl, pH 7.4, 1mM EDTA), layered on an 11.5-ml 5-20% sucrose gradient in SDS buffer (no more than 400 μg RNA in 0.5 ml are layered), and centrifuged for eight hours at 25° C at 283,000 $g_{max}$. The RNA of the appropriate sedimentation value is pooled, precipitated by ethanol and finally purified by repetition of the oligo(dT)-cellulose binding step.

A typical preparation of chorion and globin mRNA is presented in Table 3 and Figures 8 and 9.

### D. Gel Electrophoresis of RNA and cDNA

Aqueous Gels

Cylindrical (9 x 0.5 cm) or slab (17 x 17 x 0.17 cm) 6% polyacrylamide gels (20:1 acrylamide to bis) with a 4% capping gel (one fifth) are used for electrophoresis

TABLE 3

Preparation of mRNA

| RNA Type | Purification Step | Amount (μg) | Total in Each Step (%) |
|---|---|---|---|
| Chorion (about 1000 follicles) | Experiment I | | |
| | 1st oligo(dT)-cellulose[a] | | |
| | Flow-through | 1944 | 88.9 |
| | Bound | 231 | 10.6 |
| | KOH[b] | 12 | 0.5 |
| | 2nd oligo(dT)-cellulose[c] | | |
| | Flow-through | 10.7 | 31.9 |
| | Bound | 22.8 | 68.1 (mRNA, 1% initial polysomal RNA) |
| | Experiment II | | |
| | 1st oligo(dT)-cellulose | | |
| | Flow-through | 1678 | 90.6 |
| | Bound | 161 | 8.7 |
| | KOH | 14 | 0.7 |
| | 2nd oligo(dT)-cellulose | | |
| | Flow-through | 6.0 | 20.0 |
| | Bound | 24.0 | 80.0 (mRNA, 1.3% initial polysomal RNA) |
| Globin | 1st oligo(dT)-cellulose | | |
| | Flow-through | 110.80[d] | 98.53 |
| | Bound | 1.63[d] | 1.45 |
| | KOH | 0.02[d] | 0.02 |
| | 2nd oligo(dT)-cellulose | | |
| | Flow-through | 35 | 15 |
| | Bound | 197 | 85 (mRNA, 0.2% initial polysomal RNA)[e] |

[a] 1st oligo(dT)-cellulose step is on total polysomal RNA (200 mg oligo(dT)-cellulose column, Amersham).
[b] KOH refers to the removal of strongly bound RNA from the oligo(dT)-cellulose column with 0.1N KOH.
[c] 2nd oligo(dT)-cellulose column is run following fractionation on a sucrose gradient(Fig. 8).
[d] In milligrams.
[e] The low yield of pure globin mRNA may be due to aggregation.

of chorion or globin mRNA. We use slabs almost exclusively, since they permit convenient detection of the nucleic acids by $^{32}P$ autoradiography.

The slab method adopts the gel apparatus described by DeWachter and Fiers [72]. The gel is cast between two 20 x 20 cm glass plates held 1.7 mm apart with two side strips of Plexiglas (1.5 x 20 cm) joined to the glass with Vaseline. Six 2" binder clips hold the assembly together. Eight sample wells are formed at one end of the slab with a Plexiglas template, of the same thickness as the strips. There are two ways to pour the acrylamide solution into the assembly: (a) the assembly, without the template, is held vertically and sealed temporarily at the lower end with Plasticine. It is filled by pouring in the acrylamide solution at one edge, with the aid of a 10-ml pipette, and the template is inserted in the liquid before polymerization occurs. The sides of the template are coated with vacuum grease to prevent monomer seepages and a resulting formation in the sample wells of thin gel layers that collapse when the template is removed. (b) The template is inserted at once, and all potential leakage points around it are sealed with Plasticine. The assembly is then inverted, and 1-2 ml solution with excess catalyst and accelerator ("sealer") are poured in from both sides and left to polymerize for a few minutes. Then the rest of the solution is poured into the assembly and overlayered with a small volume of water.

The slab is allowed to polymerize for 1 to 2 hours. The template is removed gently and the sample wells are filled with reservoir buffer. The bottom of the slab is placed in the lower reservoir (anode) containing 0.5-1 liter of buffer. Samples are layered slowly at the bottom of each well, under the buffer, with a screw-controlled micropipet. The volume of the sample should not exceed 100 µl, and small volumes are preferable. A wick of Whatman #3 MM paper (16.8 x 16 cm, folded in half), soaked with buffer, connects the top of the gel to the upper reservoir (cathode), filled with 0.5-1 liter of buffer .

The slab gel is prepared in Tris-borate-$Na_2$EDTA (TBE) buffer [73]. A 10 X stock solution of TBE buffer, containing 108 g Tris base, 55 g boric acid, and 9.3 g $Na_2$-EDTA/liter is prepared and stored at toom temperature (0.9M Tris-borate, pH 8.3, 2.5 mM $Na_2$EDTA). If precipitates are formed after prolonged storage, they cannot be redissolved and fresh buffer should be prepared. The RNAs are electrophoresed under partially denaturing conditions, since 7M urea is included in both the gel and the sample buffer [74]. This type of "urea gel" was optimized for chorion mRNA resolution by Dr. R. Gelinas. For other types of "urea gels," see refs. 75 and 76.

To prepare a 6% slab gel with a 4% capping gel, the

following solutions are used:

| | 6% | 4% |
|---|---|---|
| Acrylamide | 3.60 g | 0.60 g |
| bis | 0.18 g | 0.03 g |
| Urea | 25.20 g | 6.30 g |
| 10x TBE | 6.0 ml | 1.5 ml |
| Water | to 60 ml | to 15 ml |

The solution is filtered, degassed, and polymerized by addition of catalyst and accelerator:

| | 6% | 4% |
|---|---|---|
| Ammonium persulfate (10%) | 0.50 ml | 0.10 ml |
| TEMED | 0.02 ml | 0.01 ml |

The gel and running buffers consist of 1 X TBE buffer. The sample buffer contains 7M urea, 0.1 X TBE buffer, 15% sucrose, 0.5M KCl, and a trace of bromophenol blue. The KCl is added because, surprisingly, it was found empirically to enhance the resolution of chorion mRNAs. The sample must be heated before layering, to minimize aggregation which leads to lack of reproducibility. Heating at $60^{\circ}$ C for three minutes and fast cooling is satisfactory.

The gel is run for 18 hours at a constant voltage (320 V). The $^{32}$P-labeled RNA is detected by autoradiography. One of the glass plates is removed, the wet gel is covered with Saran Wrap, and a Kodak X-ray film (RP-R54) is placed on top of the slab and exposed for the appropriate time.

Formamide Gels

Polyacrylamide gels containing high formamide concentrations permit electrophoresis of nucleic acids under completely denaturing conditions and thus eliminate uncertainties due to conformation or intermolecular association. In general, mobility is linearly related to the log of molecular weight (cf. SDS gels for proteins). The method was first developed by Staynov et al. [77] and further modified by Gould and Hamlyn [78,79], who used the procedure for separating the mRNAs for α and β globin. Their major modification is a buffering of the formamide at pH 9 with 0.02M diethyl barbituric acid. A full description of this method for cylindiical gels is given by Pinder et al. [81].

Maniatis et al. (in preparation) successfully used 5% polyacrylamide slab gels containing 98% formamide to resolve small DNA fragments, 50 to 1000 nucleotides long. They used phosphate as the buffer, since they found that diethyl barbituric acid gives variable results. We used this type of gel to measure the length of reverse transcripts; our experience with RNA is limited. It appears, however, that different calibration curves are obtained for RNA and DNA.

To Prepare the Gel. Deionize 100 ml 99% formamide (Matheson et al.) by stirring for 1 hr in a beaker with 5 g mixed ion exchange resin (Bio Rad AG 501-X8, 20-50 mesh). Do not stir for more than one hour, because the formamide will absorb $H_2O$ from the air and the resin. Do not leave the resin in the formamide, but filter immediately through Whatman #1 paper. This formamide must be used within 24-48 hours, or it should be deionized again.

For a 5% gel, dissolve 3.19 g acrylamide and 0.56 g bis in about 60 ml formamide (for a 4% gel, 2.55 and 0.45 g, respectively).

Bring the volume to 74 ml and add 1 ml of an aqueous solution containing 100 mg $(NH_4)_2 S_2O_8$, 170 mg $Na_2HPO_4$ and 40 mg $NaH_2PO_4 \cdot H_2O$ (final $PO_4$ concentration 0.02M, pH 7.5). Filter through Whatman #1 filter paper. Degassing is not necessary.

Add 150 μl TEMED and cast the slab by the first method described for aqueous gels. Allow the gel to "age" for at least 10 hours before use.

Remove the template and add deionized formamide in the sample well until use. Just before applying the samples, replace formamide with electrophoresis buffer.

Samples must be freed of all traces of ethanol before being dissolved. Sometimes, dissolution is not easy in deionized formamide alone. Therefore, we use the following procedure: we dissolve the samples in 10 μl $H_2O$ and add 50 μl formamide and 5 μl 0.5% each xylene cyanol and bromophenol blue. The dyes are convenient markers; the former has a mobility slightly less than half of the latter. The samples are boiled for 1 min and cooled rapidly on ice before application.

The reservoir buffer is 0.02M $Na_2HPO_4$ (pH 7.5). Upper and lower reservoir buffers must be circulated to maintain the constant pH. The gel is run at room temperature at constant voltage (200 V). For cDNA resolution, the gel is run until bromophenol blue migrates about 13 cm from the origin (about 5.5 hours).

## Gel Staining

Our experience is with methylene blue and pyronin Y. Methylene blue stains both DNA and RNA; pyronin Y is specific for RNA.

Methylene Blue Stain. Remove one glass plate from the slab gel assembly. Place the second plate with the gel facing downward in a large dish containing 5.6% acetic acid. Keep moving the plate gently, until the gel floats, and then withdraw the plate. Shake the dish for 15 minutes or for 30 minutes or more if the gel is a formamide gel.

Pour off the acetic acid solution and replace it with methylene blue stain (0.04% methylene blue in 0.2M sodium acetate, pH 4.7). Stain for one hour at room temperature, on a shaker. Destain by several changes of water.

Pyronin Y Stain

Step 1, as for methylene blue stain.

Stain overnight with 0.1% pyronin Y in 0.5% acetic acid and 1mM citric acid.

Destain by several changes of 5-10% acetic acid. The pyronin Y stain is slower than the methylene blue stain; its background is higher. Gels stained with methylene blue fade after some time, but they can be restained. Both stains can be reused several times.

Cylindrical gels or longitudinal slices of slab gels can be scanned in a spectrophotometer at 667 or 546 nm, when stained with methylene blue or pyronin Y, respectively.

Recovery of Labeled mRNA or cDNA from the Gel

Bands detected by autoradiography can be excised from the gel and the nucleic acid extracted with high salt (different authors use 0.5-0.6M ammonium, sodium, or lithium chloride, with or without EDTA and SDS). Gel material is invariably eluted together with the nucleic acid, however (see also Ref. 82); this material can be precipitated with ethanol or during dialysis, and cannot be completely removed by filtration or centrifugation. In our experience, it inhibits transcription. It may be possible to circumvent this problem by the use of ethylene diacrylate as a cross-linking agent, instead of bis, and CTAB precipitation of the extracted nucleic acid [83]. If low yields are not of primary concern, the nucleic acids can be eluted electrophoretically. Our procedure is as follows. A piece from a urea slab gel (about 1 x 1.5 cm) is placed at the bottom of a long electrophoresis tube (1 cm in diameter) and occluded in 1 ml of newly cast 6% acrylamide gel. The bottom of the tube is filled with buffer and connected to the upper reservoir through a paper wick. After electrophoresis at 320 V for 12 hr in the cold room, carrier yeast tRNA is added to the RNA eluted into the dialysis bag, and the RNA is precipitated. The yield is about 40%.

A better procedure, which leads to higher yields, is as follows. The gel piece, from a urea or formamide gel, is placed in a siliconized and autoclaved scintillation vial and smashed with a large Teflon stirring bar or (preferably) a Teflon spatula. Add 5-10 ml of 0.5M $NH_4$ acetate, 0.01M Mg acetate, and 0.1% SDS [84] and leave the vial overnight at room temperature. At this point 95% of the nucleic acid has been eluted from the gel; eventual recovery depends primarily on the size of the piece. The solution is filtered through a wet cellulose acetate filtrate filter (Schleicher and Schuell) in a Swinnex apparatus, using a plastic disposable syringe. The filtrate contains soluble material from the gel and cannot be precipitated by ethanol. If we are dealing with mRNA, we pass the filtrate through an oligo(dT)-cellulose column and wash thoroughly with the extraction buffer. The poly (A)-containing RNA binds to the column and is eluted as usual (low molarity Tris or $H_2O$). If we are dealing with cDNA, we pass the filtrate through a small (1-3 ml) poly (A)-Sepharose column [58, 85, 86]. The cDNA is eluted with 0.1N NaOH and neutralized. We never reuse this poly(A)-Sepharose column. The column-purified nucleic acids can now be precipitated by ethanol, if necessary, after addition of carrier tRNA.

## VI. TRANSLATION OF mRNA IN THE WHEAT GERM CELL-FREE SYSTEM

### A. General

Although a well-defined, reconstituted cell-free system [87] is desirable, crude lysate systems are convenient and still in general use. The most popular are the reticulocyte lysate, the Krebs ascites II, and the wheat germ systems. These systems are compared in Table 4. A brief review of the first two has been published [88]. The wheat germ system, which offers certain advantages, has found widespread use recently and is described below. Numerous eukaryotic mRNAs have been efficiently translated in this system, including those coding for globin [89, 90], immunoglobulins [91], ovalbumin [92], chorion [8], collagen [91, 92], and actin [93]. The system also translates plant [94-97] and animal [98] viral RNA, and even phage (Qβ) RNA, although at a low efficiency [99]. The wheat embryo cell-free system [100] is similar, but seems less active, and involves a somewhat tedious isolation of the embryos from seeds.

TABLE 4

Comparison of Three Cell-Free Translation Systems

| Characteristics | Reticulocyte Lysate | Krebs Ascites II | Wheat Germ |
|---|---|---|---|
| Preparation and Cost | Difficult preparation. Moderate cost (rabbits) | Cumbersome preparation. High cost(mice) | Fast, easy preparation. Low cost |
| Pre-incubation | -- | Necessary to reduce endogenous activity | Not absolutely necessary |
| Background | Very high | Low | Very low |
| Stimulation over background | Negligible Special methods (antibodies) to identify products. | Less than 20-fold unless supplemented with reticulocyte factors | Over 100-fold |
| Re-initiation | + | - | + |
| Fidelity | Good | Relatively poor | Relatively poor |
| tRNA requirement | -- | + | - |
| Stability upon storage ($-80^{o}C$) | High | Moderate | High |

B. Preparation of Wheat Germ Extracts

Commercially available raw (not toasted) wheat germ flour is used for the preparation of the extract. Not all brands are suitable; those known to be satisfactory are from "Bar-Rav" Mill (Tel Aviv, Israel), General Mills, Inc. (Minneapolis, Minn.), Old Stone Mill, Niblach's Inc., (Rochester, N.Y.), and Pillsbury's best, Pillsbury Co., (Minneapolis, Minn.). Wheat germ must be stored until use under vacuum, in a dessicator, at $4^{o}$ C.

The following procedure for the preparation of the extract is the one published by Roberts and Paterson [89], with minor modifications (Roberts and Mulligan, personal communication). Another modified procedure was recently published by Marcu and Dudock [92].

The quality of the water to be used for preparation of the solutions is crucial. It should be deionized or twice-distilled in glass and checked with a conductivity meter.

To make the extract, prepare and sterilize the following items;

Extraction Buffer (30 ml)

20 mM HEPES(pH 7.6) (adjusted with KOH)
100 mM KCl
1 mM $MgAc_2$
2 mM $CaCl_2 \cdot 2H_2O$
6 mM 2-mercaptoethanol (1% v/v mercaptoethanol is 0.143M)

Mercaptoethanol cannot be autoclaved and must be added directly to the sterilized solution.

Elution Buffer (about 2 1)

20 mM HEPES(pH 7.6)
120 mM KCl
5 mM $MgAc_2$
6 mM 2-mercaptoethanol

Record the exact volume and molarity of the potassium hydroxide used for pH adjustment, because it will be used in the calculation of the final potassium ion concentration in the reaction mixture.

Pre-incubation Master Mix (1 ml)

2.6 mM $MgAc_2$
1.0 mM ATP (neutralized with KOH)
0.2 mM GTP (neutralized with KOH)
2.0 mM DTT
8.0 mM creatine phosphate
50 units/ml creatine phosphokinase (enzyme specific activities vary, so the concentration should not be expressed as μg/ml)

Add one ml of this 10 X solution to 9 ml extract to give the above final molarities. Since $Mg^{2+}$ is already present at a concentration of 1 mM in the extraction buffer, its final concentration will be 3.5 mM.

Swell about 40 g of Sephadex G-25 (a coarse grade is fine) in elution buffer, then autoclave, cool, discard the fine particles, and pour in a column (to be about 2 x 50 cm).

Autoclave the mortar and pestle, pipettes, and tubes for the preparation and storage of the extract, a spatula, and sand (Baker). Wash thoroughly and dry the sand before sterilizing.

Work in the cold room. Grind briefly 6 g wheat germ

with 6 g of sand until a powder is formed. Add 28 ml extraction buffer and keep grinding until nothing sticks to the pestle. Transfer the homogenate to two 30-ml Corex tubes and centrifuge at 0° C, 30,000 g for 10 minutes, then remove or displace carefully with a spatula a layer of lipid formed at the surface, and transfer about two-thirds of the supernatant with a Pasteur pipette to a clean vessel.

Take 9 ml of the supernatant, add 1 ml pre-incubation master-mix and incubate for 12 minutes at 30° C. This step reduces subsequent background incorporation by about 50% and is recommended if the amount of mRNA to be tested is limiting.

Pass pre-incubated S-30 over the column at a flow rate of about 1.3 ml/min and collect 1-ml fractions of fast eluting, turbid material. A low molecular weight, bright yellow dye remains in the column. Discard the initial dilute fractions and pool only the first half of the distribution of turbid material, since the rest may be contaminated with low molecular weight material (amino acids).

Centrifuge the pooled desalted fractions at 0° C, 30,000 g for 10 minutes. This reduces the protein concentration of the S-30 by 50% without impairing its translational activity; it is important for maximizing the amount of cell-free translation mixtures which will be analyzed subsequently by electrophoresis.

Divide the supernatant of the last centrifugation (avoiding the pellet) into small tubes. These can be stored in liquid nitrogen for more than six months without loss of activity.

### C. Translation

Reaction mixtures are incubated at 25° C and contain, in a final volume of 25-50 μl, the following:

Wheat germ extract
20 mM HEPES (pH 7.0)
1 mM ATP
25-200 μM GTP
2 mM DTT
8 mM creatine phosphate
12 μM radioactive amino acid(s)
$Mg\ Ac_2$
KCl
mRNA(or polysomal RNA)

It is not necessary to add tRNAs to the system, because they are already present in the extract. The creatine phosphokinase added for the preincubation step may be sufficient, but it should be tested to see if compensating it to 50 units/ml increases the incorporation.

A preliminary optimization curve for the amount of S-30 to be added must be run for every new preparation. Usually 10 μl of extract in 50 μl of reaction mixture is optimal.

The molarity of HEPES in the reaction mixture is not crucial, but the pH is. The HEPES already present in the extract should be ignored, and new HEPES (pH 7.0) should be added to a molarity of 20 mM. (The higher pH of the extraction and elution buffers is necessary, because the extract itself is very acidic. The pH of the extract after the column and the second centrifugation is about 6.9; the extract in $H_2O$ has a pH of about 1.)

Solutions of ATP and creatine phosphate must be prepared fresh every three weeks. But it is more convenient to prepare and store a master mix in small aliquots, in liquid nitrogen.

The unlabeled amino acids must be neutralized before adding them to the master mix so that they will go into solution. The radioactive amino acids must be lyophilized before addition, if their volume is greater than 5 μl, because they contain HCl.

A final concentration of about 3.5 mM $Mg^{2+}$ is optimum for a variety of mRNAs [90,91,97] including chorion mRNA. Therefore, for a previously untested mRNA, only a narrow optimization curve (3,3.25, 3.5, 3.75, and 4.0 mM $Mg^{2+}$) is necessary. The extract is 5 mM in $Mg^{2+}$, and this source of $Mg^{2+}$ should be considered in calculating the final concentration.

The concentration of $K^+$ optimal for incorporation is around 100 mM, but varies for different mRNAs, and careful optimization is needed for each mRNA. A knowledge of incorporation dependence on $K^+$ is important in another respect: to optimize the fidelity. Although the wheat germ system at saturation levels of mRNA usually stimulates incorporation more than 100 times over background, incomplete products smaller than the in vivo synthesized protein(s) can often be seen on gels ([91]; this also occurs with chorion mRNA). These low molecular weight products ("early quitters") are suppressed if translation is done at a potassium chloride concentration somewhat higher than the one for optimal incorporation [91]. For best results, the balance between incorporation and fidelity should be optimized for each mRNA. The final concentration of potassium must be calculated carefully for it is present in the extract, in the HEPES of the extract, and in the HEPES added to the reaction mixture.

For certain RNAs (e.g., TMV) addition of 40 μM spermine (free base) increases the incorporation significantly (B. Roberts, personal communication). Marcu and Dudock [92] also reported that spermine tetrahydrochloride at a concentration of 30 μg/ml (about 80 μM) stimulated incorporation more than 1.5-fold with TMV RNA. but had no effect when chorion mRNA was used (we confirmed this finding).

A time study of the reaction is also necessary. For certain mRNAs, the reaction does not reach a plateau even after three hours of incubation. Once the other conditions are established, a saturation curve for mRNA should be obtained, if sufficient mRNA is available.

Chorion mRNAs have been translated successfully in the wheat germ cell-free system [8]. The product has been identified as chorion because it is precipitated by chorion-specific antibody, and because it exhibits the characteristic size and isoelectric point distribution of chorion. Figure 10 shows the $^{3}H$-labeled cell-free products from a wheat germ extract stimulated by the addition of partially purified chorion mRNA (0.5 μg, $Mg^{2+}$ precipitated, oligo(dT)-cellulose-bound RNA). The translation was done in the presence of 120 mM $K^{+}$ (Fig. 10A) (optimum for incorporation; 100 mM, Fig. 10B). Although the incorporation was 47% of that at optimum potassium concentration, the fidelity was improved. The products were mixed with authentic [$^{14}C$] chorion prior to electrophoresis on an SDS-polyacrylamide gel designed to fractionate small chorion-sized proteins (see Sec. II.D.2). The in vitro product did not exactly parallel the internal standard, as might be expected, in view of the multiplicity of chorion proteins. Although some of the discrepancy may be due to post-translational modifications, which do not occur in the cell-free system, much of it is probably due to an unequal translation of different messengers by the extract.

### D. Immunoprecipitation

Immunoprecipitation with specific antibody, followed by analysis of the pellet by SDS-polyacrylamide gel electrophoresis, is a good criterion for the identification and quantitation of a labeled protein (e.g., Ref. 31). The exact conditions of immunoprecipitation will vary, according to the nature of the antigen and antibody and their concentration and affinity. In each case, a standard precipitin curve must be prepared by standard procedures. For precipitating cell-free products, we mix the appropriate aliquots of translation mixture and rabbit antibody (in slight excess) with buffer (0.1M phosphate, pH 7.5, 0.15M NaCl, 1% Triton X-100 to minimize nonspecific precipitation), incubate overnight at $4^{o}$ C, centrifuge, wash the pellet two times with the above buffer, and dissolve it in the electrophoresis sample buffer.

If the antibody is very weak, it may be necessary to use a double-antibody procedure, in which the products are allowed to react with rabbit antibody and the complexes are precipitated with a slight excess of goat anti-rabbit serum (P-4, Antibodies, Inc., Davies, Calif.). We usually find a ratio of 2.5 for goat serum: rabbit serum appropriate. Care should be taken to use only as much serum (or sera) as is warranted by the amount of antigen present, in order to minimize nonspecific precipitation.

## VII. REVERSE TRANSCRIPTION

### A. General

The RNA viruses contain an RNA-directed DNA polymerase (reverse transcriptase); for an excellent and extensive review see Green and Gerard [101]. Reverse transcription, however, is not a unique property of these enzymes, since it has been demonstrated that other polymerases, including DNA polymerase I of *Escherichia coli* can, under certain in vitro conditions, transcribe a primed RNA template [102-105]. On the other hand, the viral reverse transcriptases can catalyze a repair-like reaction at a primed single-stranded DNA region.

In this section we will examine the technical details of a procedure using avian myeloblastosis virus (AMV) reverse trascriptase(RT) to transcribe eukaryotic mRNAs to cDNA. (For the various uses of cDNA, see P. Williamson, this volume.) AMV-RT is widely used, not only because of its greater transcriptional efficiency as compared with the easily available *E. coli* polymerase I, but also because of a substantial (although not complete) body of information in the literature concerning its use. A good selection of papers for the beginner can be found in the literature [106-116].

AMV-RT, although now available commercially (Boehringer Mannheim Co.), is usually purified from the virus.

The virus can be obtained from Dr. J. W. Beard (Life Sciences, Inc., Gulfport, Fla.), but only after making the proper arrangements with Dr. M. A. Chirigos (National Cancer Institute, Bethesda, Maryland); we are grateful to both for a generous supply of AMV.

### B. Purification of AMV-RT

We purify the enzyme according to the procedure of Verma and Baltimore [107], from AMV present in the plasma of leukemic chickens. The virions are purified, the polymerase is solubilized by a nonionic detergent, and the active fraction is isolated by successive DEAE-Sephadex and phosphocellulose chromatography. By this procedure the enzyme is only 50- to 100-fold purified, but it is pure enough for transcription of eukaryotic mRNA. If greater purity is desired, the purification scheme of Leis and Hurwitz [108] can be used. Also, AMV-RT has been purified to more than 90% purity by two other procedures [110, 127].

Although there are no reported effects of AMV on humans, all precautions should be taken during the handling of the leukemic plasma (use of gloves , no pipetting by mouth, disposal of debris in base, sterilization of glassware after use, etc.). Purification of RT in a laboratory that uses chickens as experimental animals should be avoided.

Two species of AMV-RT, separable by phosphocellulose chromatography, have been identified; a minor one (10 - 20% of the activity), and a major one with two subunits, $\alpha$ and $\beta$ (about 65,000 and 110,000 daltons, respectively). There are indications that the minor species is the subunit of the major, generated during the chromatographic procedure. Both species also exhibit ribonuclease H activity, but their mode of action is different; further references on the RNase H activity are available [101]. The preparative procedure calls for collection of the major peak.

The purification process should be started only when the chromatography columns are ready; it should be recalled that DEAE-Sephadex, in particular, requires a long time to equilibrate completely. Assays at various stages of purification, using a ribo-copolymer template, should be run under sterile conditions,as should transcriptions of natural mRNAs. As a final preparative step, before storage, the volume of the pooled fractions from the phosphocellulose column can be reduced by dialysis against polyethylene glycol. Substantial losses can occur, however, so this step should be omitted, unless the enzyme is extremely dilute. Since the enzyme is inactivated by repeated freezing and thawing, storage in small aliquots at $-70^{\circ}$ C is recommended. We customarily use 15 μl of enzyme preparation per 50 μl of incubation mixture. Therefore, we store the enzyme in liquid nitrogen in 20-μl aliquots.

Some preparations have slight ribo- and deoxyribonuclease activity. Therefore, each preparation should be tested for nuclease activities before aliquoting and storage. Any kind of labeled RNA and DNA can be used for this purpose, and it must be incubated in the presence of enzyme under conditions of reverse transcription. Preparations causing loss of TCA-precipitable radioactivity should be further purified by rechromatography on phosphocellulose [108] or by CM-sephadex chromatography [110]. More sophisticated assays for nuclease contamination, such as appearance of nucleic acid fragments of reduced electrophoretic mobility, are sometimes necessary. Using formamide slab gels we found endoribonuclease (but not endodeoxyribonuclease) contaminants, which had escaped detection by TCA precipitation, in a number of preparations. This activity was effectively removed by a small (0.7 x 1.5 cm) phosphocellulose column, which also concentrates the enzyme preparation. Surprisingly, we saw no difference in the DNA product length, with pure or contaminated RT.

Recently, Marcus et al. [128] reported a novel one-step purification procedure for AMV-RT: disrupted virions are passed through a small column of poly(rC)-Sepharose, which specifically binds AMV-RT. The enzyme is then eluted with a linear KCl gradient; it is 90% pure and free of nuclease.

C. Conditions of Transcription with AMV-RT

The composition of the enzyme storage buffer should be considered in optimization experiments, since it affects the composition of the transcription reaction mixture. The enzyme can be stored in the buffer used for elution from the last column, or pooled and concentrated fractions can be dialyzed against a buffer of choice. We store our enzyme directly as the eluate from the second phosphocellulose column (in 50 mM Tris, pH 7.2, 0.1 mM $Na_2$ EDTA, 0.1M 2-mercaptoethanol, 0.48M KCl, 0.2% Nonidet P-40, and 25% glycerol). With our enzyme preparation, optimized conditions for rabbit globin reverse transcription are shown in Table 5.

Temperature and pH Optima

The endogenous reaction (synthesis of DNA on viral RNA, using disrupted virions in the presence of substrate and ions) has a broad temperature optimum, centering about 40-45° C for avian viruses and 37-40° C for mammalian viruses [106]; the difference presumably reflects the higher body temperature of birds, as compared with mammals. In the case of Rous sarcoma virus (RSV), an avian virus, the optimum was at about 45-46° C for short incubations (1 hr), but about 40° C for prolonged incubations of three hours [129]. No data were shown for AMV-RT, but similar behavior was reported. The time-course of the reaction at different temperatures has been investigated for murine sarcoma virus, a rodent virus [130]. The temperature optimum for in vitro reactions of AMV-RT (using exogenous templates) has not been reported, but the reaction is routinely run at 37° C (Table 5). Using poly(rC)·oligo(dG) we compared the reaction rates at 37°, 42°, and 45° C. Although no substantial differences were observed, 42° C seems to be the optimum temperature, given the conditions of our procedure.

There is general agreement in the literature [108, 109] concerning the pH optimum. The in vitro reaction has a broad optimum between 7.8 and 8.5 or 9, with maximum activity at 8.2. The activity falls off rapidly below 7.5 and less rapidly above 8.5. At pH 7.2 and 9.8, incorporation is 50% of that at pH 8.2. With few exceptions, 50 mM Tris-HCl (pH 8.3) is used in the reaction (Table 5). Since the pH of Tris changes with temperature, the buffer should be prepared so as to maintain a pH of 8.2-8.3 at the temperature of incubation.

Substrate

The reaction is dependent on the presence of all four dNTPs. The omission of one reduces DNA synthesis markedly. However, the dNTPs have been used at different concentrations with different yields and in no case was their concentration optimized (Table 5). Usually, three

TABLE 5

Reverse Transcription Conditions of Various Eukaryotic mRNAs Using AMV Reverse Transcriptase[a]

| | mRNA μg/ml | Tris-HCl mM (pH) | dNTPs[b] | $MgCl_2$ or Mg acetate (mM) | NaCl or KCl (mM) | DTT (or mercapto-ethanol, ME) (mM) |
|---|---|---|---|---|---|---|
| 10S rabbit globin [113] | ~3 | 50 (8.3) | 1<br>160 (1) | 6 | 60 (Na) | 10 |
| 10S rabbit or human globin [114] | 20 | 50 (8.3) | 0.2<br>40 (1) | 6 | 50 (K) | -- |
| 9S rabbit globin [115] | 1 | 100 (7.2) | 0.2<br>70 (1) | 6 | 40 (K) | 2 |
| [117] | 60 | 50 (8.3) | 0.4 (2)<br>400 (2) | 10 | 140 (K) | 30 (ME) |
| 9S mouse globin [118] | 5-10 | 50 (8.2) | 0.5<br>40 (2) | 5 | 50 (K) | 10 |
| 9S duck globin [119] | 10 | 50 (8.3) | 1<br>40-50 μM (2) or 3.5-4 mM (2) | 10 | 10 (K) | 1.5 (ME) |
| 10S duck globin [120] | 6 | 50 (8.3) | 1<br>800 (1) | 6 | 60 (Na) | 20 |

| | | | | | | |
|---|---|---|---|---|---|---|
| pigeon globin [121] | 10 | 50 (8.3) | 2<br>30(1) | 6 | 60 (K) | 4 |
| calf lens α-crystallin 14S [116] | 6 | 50 (8.3) | 0.6<br>80 (1) | 6 | 60 (Na) | 20 |
| 10S or 14S [122] | 20 | 50 (8.3) | 0.2<br>40 (1) | 6 | 50 (K) | -- |
| ovalbumin 17-19S [123] | 11 | 50 (8.3) | 0.025<br>21(3) | 6 | 30 (Na) | 10 |
| [124] | 8 | HEPES (7.5) | 0.2<br>9 | 6 | 60 (K) | 10 |
| 14S IgG light chain [125] | 10 | 50 (7.5) | 1<br>5 (1) | ? | 10 (K) | 1.5 (ME) |
| Fibroin (5.5-6x10$^{6}$ daltons) [126] | 5.5 | 50 (8.25) | 0.5<br>25 (1) | 6 | 55 (Na) | 10 |
| Dictyostelium RNAs (avg. 1000 bases) [144] | 1.5 | 50 (8.3) | 0.6<br>100 | 6 | 60 (Na) | 10 |

TABLE 5 (Cont'd.)

| | mRNA µg/ml | Tris-HCl mM (pH) | dNTPs[b] | $MgCl_2$ or Mg acetate (mM) | NaCl or KCl (mM) | DTT (or mercapto-ethanol, ME) (mM) |
|---|---|---|---|---|---|---|
| | Oligo (dT) (µg/ml) | Actinomycin (µg/ml) | Catalase or BSA (µg/ml) | Incubation (min at 37° C) | Yield[c] | Length |
| 10S rabbit globin [113] | 0.05 | 20 | -- | 90 | 30-80% | 8S |
| 10S rabbit or human globin [114] | 2 | 100 | -- | 30 | ~3% | 8.3S |
| 9S rabbit globin [115] | 10 | -- | -- | 60 | ~7-11% | 6.3,5.8S |
| [117] | 20 | -- | -- | 60(40°) | 14.5% | mainly 650 bases |
| 9S mouse globin [118] | 1-2 | 20 | 100 | 90 | 20% | 330 bases |
| duck globin 9S [119] | 2 | 35 | -- | 30 | 6% or 50% | 5.7S or 6.9S |
| [120] | 0.2 | 30 | -- | 90 | ? | 7-8S |
| pigeon globin [121] | 1.25 | 20 | -- | 60 | ~4% | 7.8S,6.5S |
| calf lens - crystallin [116] | 10 | 100 | -- | 180 | 20-60% | <8S |
| [122] | 4 | 100 | -- | 50 | ~0.15-0.5% | 7.6S (10S) or 8.3S (14S) |

| | | | | | | |
|---|---|---|---|---|---|---|
| ovalbumin 17-19S [123] | 5 | 20 | 100 | 60 (20°) | ? | 5.2S |
| [124] | 5 | 20 | 1 mg/ml | 90 | ? | 200 bases |
| IgG light-chain [125] | 2 | 25 | -- | 20-30 | ? | 5S |
| fibroin (5.5-6x10[6] daltons) [126] | 0.5 | -- | 80 | 120 | 7% | 7S (4-11S) |
| dictyostelium RNAs (avg. 1000 bases) [144] | 10 | 100 | -- | 120 | 62% | 500-600 bases |

[a]Amount of enzyme used not listed due to the difficulty of comparison of units used from lab to lab.

[b]Number in parenthesis denotes number of labeled dNTPS used; unlabeled dNTPs in mM, labeled dNTPs in μM.

[c]Percent of the input template transcribed into cDNA(reported or calculated).

of the dNTPS are unlabeled, and the fourth, labeled dNTP, is used at a lower concentration for reasons of economy and higher specific activity (this nucleotide should not be dTTP if the cDNA is to be used for hybridization, P. Williamson, this volume). It has never been shown that the use of unlabeled dNTPs at higher concentrations is beneficial. In fact, at least for the overall rate of the reaction, the concentration of the labeled dNTP was shown to be the limiting factor in an endogenous reaction with RSV-RT [131]: increasing the concentration of the labeled dNTP from one to 500 mM (with the unlabeled dNTPs in excess) led to a proportional increase in the rate of DNA synthesis.

In in vitro reactions with AMV-RT, the influence of dNTP concentration on yield and product length was either overlooked or not reported, except in two cases. One publication [119] reports that, with the unlabeled dNTPs held constant at 1 mM, duck globin 9S mRNA was transcribed with a 6% yield and to an average product size of 5.7S when the labeled dNTPs were present at 40-50 μM, and with a 50% yield and to an average product size of 6.9S when the labeled dNTPs were raised to 3.5-4 mM. According to another report, when unlabeled dNTPs were maintained at 1 mM a heterogeneous group of poly(A)-containing RNA from mouse myeloma tumor (MOPC-41), with a main peak of 14S, was transcribed with the labeled dNTP at concentrations of 4, 40, and 400 μM to average sizes of 5.9, 6.3, and 7.6S, respectively [132]. The accuracy of the measurement in both these reports was limited, since the size was estimated by alkaline sucrose gradients and with BSA run as a marker in parallel neutral gradients. For length measurement, it would be desirable to use isokinetic gradients [133] with an internal standard (labeled DNA restriction fragments of known length are ideal). Alternatively, a linear gradient can be used if the sample is bracketed by two standards. A far superior method for measuring lengths is electrophoresis on formamide slab gels, in parallel with restriction fragment markers.

We have explored the influenece of dNTP concentration by transcribing globin mRNA in the presence of dNTPs at concentrations varying from 5 to 400 μM [117]. In our initial experiment, the increase in product size with concentration of all four dNTPs was determined by sucrose gradient analysis, using internal marker DNA. For further work [$^{32}$P]cDNAs were analyzed on formamide slab gels and detected by autoradiography.

From the outset, two unexpected results were obvious. The reverse transcripts fell into discrete size classes, rather than into a continuous spectrum of sizes; and the largest transcripts were of the same size as globin mRNA (650 nucleotides), suggesting that they were complete copies. The increasing average size of the transcripts with increasing dNTP concentration was seen to result from a shift in the number of the various transcript size classes. As the concentration of all four dNTPs increased, the full-size cDNA increased at the expense of the incomplete transcripts. Lowering the concentration of even a single nucleotide was sufficient to shift the balance toward incomplete transcripts. The discrete nature of the incomplete transcripts suggests that they are generated beaause of the presence of mRNA regions that are difficult to transcribe, which are possibly regions of substantial secondary structure.

Figure 11 shows a typical experiment in which nucleotide concentrations were varied. A dramatic contrast is afforded by the reverse transcripts synthesized at extremely low and at moderately high nucleotide concentrations (each nucleotide at 5 μM or at 100 μM, respectively; "all-5" and "all-100" samples). In the former, only trace amounts of full-sized product are present, and the predominant components are very small (<140 NT). In the latter, full-sized product predominates, and among the incomplete fragemnts the larger ones predominate. With one of the substrates kept at 5 μM, and when the concentration of the remaining three was increased to either 100 μM or to 500 μM, a profile intermediate between those of the "all-5" and "all-100" samples was seen (Fig. 11, "5/100" and "5/500" slots): discrete components ranging from 60 NT to full size were seen in equal numbers. Increasing the concentration of the least abundant nucleotide from 5 to 25 μM shifts the pattern substantially toward that of the "all-100" sample (cf. slots "all-100," "5/100," and "25/100" in Fig. 11).

Low nucleotide concentrations also greatly decrease the yield of cDNA. In terms of both yield and number of full-size copies, dNTP concentrations lower than 25 μM were particularly inadequate in our procedure. Above the critical level, steady but gradual improvement was noted; concentrations as high as 400 μM did not appear to be saturating.

All fragments displayed in Figure 11 are DNA: they are TCA-and ethanol-precipitable, alkali-resistant, sensitive to S1 nuclease, and dependent for their formation on the complete reaction mixture, including the globin template, the oligo(dT) primer, and the enzyme. Additional $^{32}$P-containing components are generated during the reaction; they are ethanol-precipitable but TCA-soluble. They can be separated easily from the bona fide transcripts, since they are extensively retarded on a Sephadex G-100 column; they are presumably polyphosphates.

It was also observed that the nature of the limiting nucleotide affects the pattern of incomplete transcription. Reverse transcripts were obtained from reactions in which either [$^{32}$P]dATP or [$^{32}$P]dGTP concentration were limiting. Electrophoresis revealed many incomplete fragments of similar size in the two samples; but significant differences were also apparent.

With moderately high dNTP concentrations (all 100 μM) chorion mRNA yields transcripts in discrete size classes, ranging up to the size of the respective template (Fig. 11). The same holds true for ovalbumin mRNA, which is between 1670 and 2640 nucleotides long [124] (data not shown). We conclude that reverse transcriptions should be performed with all four nucleotides at high concentrations. The exact concentration should be chosen keeping in mind the conflicting goals of high yield and full sequence representation, on the one hand, and high specific activity without inordinate expense, on the other. If desired, full-length cDNA can be recovered after fractionation on the formamide gels.

Template and Primer

The reaction requires a template, and a primer with a free 3'-OH group. A wide variety of eukaryotic mRNAs can be transcribed (Table 5), although primed homopolymers are better templates. The efficiency of transcription, under fixed conditions, depends on the template (but unexplained variability is also observed in repeated experiments). In pilot experiments with a new mRNA, saturation levels of the template for a given amount of enzyme should be determined, provided sufficient mRNA is available. In our hands, for example, 60 μg/ml globin mRNA nearly saturates 15 μl of our enzyme preparation in a total reaction mixture volume of 50 μl.

For eukaryotic mRNAs containing poly(A) at the 3' end of the molecule, oligo$(dT)_{10}$ or oligo$(dT)_{12-18}$ is used as primer. Longer oligo(dT) chains are less effective primers [125]. Verma et al. [113] calculated that about two primer molecules $(dT_{10})$ per template molecule (globin mRNA) maximized transcription; additional primer has no effect. Diggelmann et al. [125]reported that about 2 ng/ml of oligo(dT) primer (of unknown length) is optimum for 10 μg/ml of 14S IgG light-chain mRNA; a large excess of this primer had a slight inhibitory effect. They reported, also, that addition of fresh enzyme 15 minutes after the beginning of the reaction increased the final incorporation about three-fold. They believed that this is due to inactivation of the enzyme. By contrast, Leis and Hurwitz [111] showed that addition of more enzyme 20 min after the beginning of the reaction does not increase the transcription of AMV-RNA, although addition of more RNA increases the incorporation two-fold.

Cations, Reducing Agents, and Protective Proteins

RT has an absolute requirement for $Mg^{2+}$ or $Mn^{2+}$. We observed that $Zn^{2+}$ has no effect on the reaction (although the enzyme is a Zinc metallo-enzyme [134]). Optimal activity under our conditions occurs at 10 mM $Mg^{2+}$, in agreement with the results of Leis and Hurwitz [111] who used

AMV-RNA as template. Other authors used lower concentrations of $Mg^{2+}$, usually 6 mM (Table 5). Leis and Hurwitz [111] reported, also, that, for AMV-RNA, the requirement for $Mg^{2+}$ could be partially met by $Mn^{2+}$ (1 mM), but, when both were present ($Mn^{2+}$ at 0.2 mM), a synergistic effect was observed (2.5-to three-fold increase in incorporation). Using homopolymers [poly(rC)oligo(dG) and poly(rA)oligo (dT)] we were unable to confirm this observation. The two ions were also not synergistic for RSV-RT with RSV-RNA and rA·dT as templates [135].

Monovalent cations ($K^+$ or $Na^+$) are not required for activity, but they can increase incorporation with certain templates. Optima must be sought for each mRNA.

The presence of a sulfhydryl reagent (DTT or 2-mercaptoethanol) is required for optimal activity, but there is no general agreement about its concentration. In our hands, 30 mM 2-mercaptoethanol is optimal; it is provided by the enzyme preparation, and addition of DTT is inhibitory.

Harrison et al. [136] reported that the addition of a protein (100 μg/ml catalase or BSA) stabilizes the enzyme, resulting in higher yields and length of transcript, especially when the substrate is of high specific activity. Other authors also use BSA, even at higher concentrations (Table 5).

Actinomycin D

Actinomycin D is used in the reaction to prevent double-stranded DNA synthesis. Various authors have used the antibiotic at different concentrations (Table 5), but control experiments showing degree of inhibition vs. concentration of the drug are usually not reported. The effect of the ethanol often used for dissolving the actinomycin stock is usually either not evaluated or not reported. The most careful study concerning the actinomycin effect was reported by Ruprecht et al. [112]. They used 100-200 μg/ml of the drug and AMV-RNA and a series of homopolymers as templates.

For preparation of stock solutions, we dissolve actinomycin in water by prolonged vortexing. At the high template concentrations we used (60 μg/ml globin mRNA), drug concentrations as high as 100 μg/ml produced a 24% inhibition. At lower template concentrations (20 μg/ml), inhibition was 27% at 300 μg/ml actinomycin (labeled and unlabeled dNTPs at 5 and 100 μM, respectively).

D. Hybridization

The micromethod described below can be used for back-hybridization of cDNA to its template (and also for studies of DNA reassociation kinetics).

Samples of cDNA (1-2 μl in $H_2O$) are introduced with a Hamilton syringe into 50 μl of mineral oil (paraffin oil, Baker) in siliconized microanalysis tubes (total capacity

250 μl; Werthemann and Co., Basel, Switzerland). The oil has been degassed and equilibrated (at the appropriate temperature) with the hybridization buffer. This guarantees that the volume of the hybridization mixture will not change during incubation (as confirmed by refractive index measurements). The tubes are heated to 97$^{\circ}$ C for 15 minutes and transferred immediately, on ice, in the cold room. Add 1-2 μl (equal volume to cDNAs) RNA in 8x PIPES buffer (10x PIPES buffer: 1.75M NaCl, 0.1M PIPES, pH 6.7) [137] to the oil, being careful not to disturb the drop of the cDNA. The tubes are centrifuged at 0$^{\circ}$ C for 15 sec to combine the two aqueous drops, and the hybridization reaction is initiated by placing the tubes at 60$^{\circ}$ C, except for two tubes, which are instantly frozen in an acetone-dry ice bath for determination of zero time double-stranded material. At selected time intervals, tubes are removed and frozen as above, and then stored in liquid nitrogen for later analysis.

For assay, the samples are thawed, and immediately 50 μl of buffer (0.2M NaCl, 0.05M NaAc, pH 4.5, 0.001M $ZnCl_2$), containing 20-50 μg/ml sonicated and heat-denatured calf thymus DNA and S1 nuclease in excess, is layered on the oil with a Hamilton syringe. The tubes are centrifuged for 15 sec to combine the aqueous phases and incubated at 45$^{\circ}$ C for 1 hr. After incubation, the aqueous phase, inevitably contaminated with oil, is transferred with a microcap onto parafilm, stretched on ice. The oil wets the parafilm, while the aqueous phase floats. Take up 50 μl of the aqueous phase into a clean microcap and pipette onto a dry 2.4 cm Whatman 3 MM filter disc, previously coated with carrier BSA. The disc (still wet) is dropped after 30 sec into cold 10% TCA. The discs are washed batchwise in TCA containing 1% sodium pyrophosphate (several changes), washed briefly with ethanol, dried, and counted after addition of tuluene-Liquifluor. If the cDNA is $^3$H labeled, the discs are incubated for 1 hr in 0.3 ml Protosol (which solubilizes the product, increasing the efficiency of $^3$H counting); then the fluor added should contain 0.3% HAc.

This micromethod, utilizing an accurate syringe and microanalysis tubes, is superior to that utilizing sealed capillaries in terms of accuracy and economy. The capillary method (because of volume variability) necessitates the splitting of the sample into two halves after annealing, for assay of total counts and counts in double-stranded (e.g., S1-resistant) material. This introduces unnecessary pipetting errors, increases the number of samples required, and consumes expensive cDNA and mRNA.

### E. S1 Single-Strand Specific Nuclease

#### Purification

As source of enzyme one can use "Takadiastase" (Sankyo Co., Tokyo) or the equivalent, crude α-amylase powder from Aspergillus oryzae (Sigma).

Any one of four purification schemes can be used [137-140]. The procedure described in Ref. 140 yields 90% pure S1 nuclease, whereas the rest yield partially purified nuclease, which is contaminated with T-1 ribonuclease but suitable for measuring both DNA/DNA and DNA/RNA annealing. The enzyme is a single polypeptide chain with a molecular weight of 32,000-36,000. Our experience is with partially purified S1 nuclease by the method of Britten et al. [137]. Stored at $-20^{\circ}$ C in 50% glycerol, it retains its full activity for more than a year. The enzyme is also commercially available (Miles Laboratories, Inc.).

Properties

The enzyme degrades single-stranded nucleic acids (both RNA and DNA, including circular single-stranded DNA), has both endo- and exonucleolytic activities (3' and 5'), and does not show sequence specificity. The endonucleolytic activity is predominant at the early stages of the reaction, whereas exonucleolytic activity predominates at later stages [139]. Denatured DNA is finally digested to 5'-deoxynucleotides [138]. The DNase/RNase activity is about 5:1 [140].

The S1 is most active at pH 4.3, with half-maximal rates at pH 3.3 and 4.9 [140]. In assays, a pH of 4.5 is generally used, to reduce nonenzymatic strand scission at low pH values.

The enzymatic activity at 45° C is twice that at 35° C [140]. The enzyme is still active at temperatures up to 60-65° C. Lower temperature incubations (37°C) should be used for assaying reassociation of repetitive DNA, since mismatching must then be tolerated.

The S1 is stimulated by zinc. Addition of EDTA to 1mM completely inactivates the nuclease. Addition of zinc to 1 mM, in the presence of 0.1 mM EDTA, restores at least 70% of the original activity. The enzyme is fully active over a broad range of zinc concentrations (0.03-3 mM) [142]. Therefore, the presence of zinc in the assay buffer at a concentration of at least 1 mM is recommended, because this eliminates the need of controls that would otherwise be necessary if small concentrations of EDTA or citrate were used in hybridization buffers, or if they occurred in the reaction mixuure as contaminants. The enzyme is significantly inhibited by sodium phosphate (pH 4.6) at concentrations as low as 10 mM, but it tolerates SDS up to 0.04%,0.8M urea, and 5% formamide [140]. Reportedly, it also tolerates 10M urea (P. O'Farrell, personal communication).

The ionic strength is an important determinant for enzymatic activity. The S-1 is optimally active at 0.1M NaCl [139, 140] , whereas, at 0.3M NaCl, there is up to 4% nuclease resistance [140,142]. For assaying reannealing of repetitious DNA, NaCl concentrations of 0.17-0.3M should be used, to tolerate mismatching. Concentrations higher than 0.4M are inhibitory. At NaCl concentrations between 3 and 300 mM, the reaction is usually allowed to proceed for 30 minutes at 50° C [139, 141] and 60 minutes at 37° C [142].

At low salt concentrations (10 mM NaCl), S-1 does not completely hydrolyze the denatured DNA present at concentrations lower than 5 μg/ml [139], a phenomenon that cannot be reversed by adding excess enzyme. Therefore, in assays, carrier heat-denatured and sheared DNA at a concentration of 10-20 μg/ml is routinely used. The role of the added carrier DNA has not been established. This cannot be replaced by double-stranded or unsheared denatured DNA [141]. The addition, however, of NaCl to 0.3M eliminates the requirement of S-1 for a critical amount of substrate [142]. This higher concentration of salt offers the further advantage of preventing hydrolysis of hybrids by contaminating T-1 nuclease.

The purified enzyme enhibits no enzymatic activity on native DNA [140]. But, some preparations of partially purified enzyme can degrade double-stranded DNA at low ionic strength to a certain extent [141, 142]. This activity on native material almost disappears at 0.3M NaCl.

Pure or partially purified enzyme in 75 or 250 mM NaCl cleaves both strands of circular,covalently closed, superhelical SV40 DNA (and also polyoma DNA)to unit-length linear duplex molecules with intact strands. (At 10 mM

NaCl, however, shorter than unit length products are generated.) Circular, covalently closed, non-superhelical DNA is resistant to the enzyme [143]. The S-1 converts RFI DNA of fd phage to RF II without further degradation [142]. At low salt concentrations (50 mM), the enzyme introduced nicks to duplexes, which substantially diminished at 0.3M NaCl [140].

## ACKNOWLEDGMENTS

We thank our friends and colleagues (in alphabetical order) R.E. Gelinas, M.R. Goldsmith, J. R. Hunsley, M. Nadel, R. Palmiter, M. Paul, W. H. Petri, J. C. Regier, N. Rosenthal, J. N. Vournakis, and A. Wyman for various contributions to this chapter, T. Maniatis and A. Jeffrey for our enjoyable collaboration on reverse transcription, a summary of which appears here, A. Maxam for considerable advice on techniques, G. Crouse and A. Frischauf for their help in RT purification, and B. Roberts and R. Mulligan who made possible the brief review on the wheat germ cell-free system. The expert secretarial assistance of L. Lawton and the excellent technical help of M. Koehler and L. DeLong are gratefully acknowledged. This chapter originated partly as a workshop in the Embryology Course at the Marine Biological Laboratory, Woods Hole, during the summers of 1972-1974; FCK thanks E. H. Davidson and his other colleagues for the stimulating environment created there. The work on the chorion has been supported by the NIH (5-R01-HD04701), the NSF (GB-35608X), and the Rockefeller Foundation (RF-73019).

## REFERENCES

1. F. C. Kafatos, J. Regier, G. D. Mazur, M. R. Nadel, H. Blau, W. H. Petri, A. R. Wyman, R. E. Gelinas, P. B. Moore, M. Paul, A. Efstratiadis, J. Vournakis, M. R. Goldsmith, J. Hunsley, B. Baker, and J. Nardi, in _Biochemical Differentiation of Insect Glands_, Vol. X of _Results and Problems in Cell Differentiation_, W. Beermann, ed. Springer-Verlag, Berlin , 1975.
2. J. F. Morrow, S. N. Cohen, A. C. Y. Chang, H. W. Boyer, H. M. Goodman, and R. B. Helling, _Proc. Natl. Acad. Sci. USA_, _71_, 1743 (1974).
3. M. Thomas, J. R. Cameron, and R. W. Davis, _Proc. Natl. Acad. Sci. USA_, _71_, 4579 (1974).
4. Committee on Recombinant DNA Molecules, Letter to _Science_, _185_, 303 (1974).
5. M. Paul, M. R. Goldsmith, J. R. Hunsley, and F. C. Kafatos, _J. Cell Biol._, _55_, 653 (1972).
6. M. Paul and F. C. Kafatos, _Dev. Biol._, (1975). In press.
7. R. E. Gelinas and F. C. Kafatos, _Proc. Natl. Acad. Sci. USA_, _70_, 3764 (1973).
8. J. R. Hunsley, R. E. Gelinas, and F. C. Kafatos, in preparation (1975).

9. R. E. Gelinas, Ph.D. thesis, Harvard University. Cambridge, Mass. 1974.

10. J. N. Vournakis, R. E. Gelinas, and F. C. Kafatos, Cell, 3, 265 (1974).

11. W. H. Petri, A. Wyman, and F. C. Kafatos, in preparation (1975).

12. W. H. Petri, A. R. Wyman, and F. C. Kafatos, in The Genetics and Biology of Drosophila, T. Wright and M. Ashburner, eds. Vol. II, Academic Press, London, in press.

13. H. Kawasaki, H. Sato, and M. Suzuki, J. Insect Biochem 1, 130 (1971).

14. P. O'Farrell, J. Biol. Chem., in press (1975).

15. O. Vesterberg, Acta Chem. Scand., 23, 2653 (1969).

16. O. Vesterberg, Methods in Enzymology, XXII, 389, (1971).

17. N. Catsimpoolas (ed.), Isoelectric Focusing and Isotachophoresis, Annals of the N. Y. Acad. Sci., 209, (1973).

18. S. N. Vinogradov, S. Lowenkron, M. R. Andonian, J. Bagshaw, K. Gelgenhauer, and S. J. Pak, Biochem. Biophys. Res. Commun., 54, 501 (1973).

19. H. Svensson, in Protides of the Biological Fluids, H. Peeters, ed. Vol. 15, Elsevier, Amsterdam, The Netherlands, 1968, p. 515.

20. J. V. Maizel, in Methods in Virology (K. Maramorosh and H. Koprowski, eds., Vol. V, Academic Press, New York, 1971, p. 179.

21. W. M. Bonner and R.A. Laskey, Eur. J. Biochem., 46. 83 (1974).

22. H. R. Maurer, Disc Electrophoresis and Related Techniques of Poly-acrylamide Gel Electrophoresis, 2nd ed., Walter de Gruyter, Berlin, New York, 1971.

23. K. Weber, J. R. Pringle, and M. Osborn, in Methods in Enzymology, C. H. W. Hirs and S. N. Timasheff, eds., Vol. XXVI, Academic Press, New York, 1972, p. 3.

24. A. L. Shapiro, E. Viñuela, and J. V. Maizel, Jr., Biochem. Biophys. Res. Commun., 28, 815 (1967).

25. J. Zwaan, Anal. Biochem., 21, 155 (1967).

26. R. T. Swank and K. D. Munkres, Anal. Biochem., 39, 462 (1971).

27. L. Ornstein, Ann. N. Y. Acad. Sci., 121, 321 (1964).

28. B. J. Davis, Ann. N. Y. Acad. Sci., 121, 404 (1964).

29. U. K. Laemmli, Nature, 227, 680 (1970).

30. W. F. Studier, J. Mol. Biol., 79, 237 (1973).

31. E. Berger and F. C. Kafatos, Dev. Biol., 25, 377 (1971).

32. F. C. Kafatos, in Current Topics in Developmental Biology 7, A. A. Moscona and A. Monroy, eds., Academic Press, New York, 1972.

33. M. A. Yund, F. C. Kafatos, and J. C. Regier, Dev. Biol., 33, 362 (1973).

34. B. S. Katzenellenbogen and L. B. Williams, Proc. Natl. Acad. Sci. USA, 71, 1281 (1974).

35. M. A. Yung, W. F. Yund, and F. C. Kafatos, Biochem. Biophys. Res. Commun., 43, 717 (1971).

36. S. Weisberg, Anal. Biochem., 61, 328 (1974).

37. L. D. Smith and R. E. Ecker, Current Topics in Developmental Biology, 5, A. A. Moscona and A. Monroy, eds. Academic Press, New York, 1970, p. 1.

38. J. Regier and F. C. Kafatos, J. Biol. Chem., 246, 6480 (1971).

39. M. B. Peterson, A. G. Ferguson, and M. Lesch, J. Mol. Cell Cardiology, 5, 547 (1973).

40. Y. Saito and J. R. Florini, Anal. Biochem., 54, 266 (1973).

41. J. C. Regier and F. C. Kafatos, in preparation.

42. F. C. Greenwood, in Principles of Competitive Protein-Binding Assays, Odell and Daughaday, eds., Lippincott, Philadelphia, 1971, p. 288.

43. P. J. G. Butler, J. I. Harris, B. S. Hartley, and R. Leberman, Biochem. J., 112, 679 (1969).

44. D. N. Talbot and D. A. Yphantis, Anal. Biochem., 44 246 (1971).

45. M. Inoye, J. Biol. Chem., 246, 4834 (1971).

46. P. Bohlen, S. Stein, W. Dairman, and S. Udenfried, Arch. Biochem. Biophys., 155, 213 (1973).

47. G. E. Means and R. E. Feeney, Chemical Modification of Proteins, Holden-Day, Inc., San Francisco, 1971.

48. G. E. Means and R. E. Feeney, Biochemistry, 7, 2192 (1968).
49. R. H. Rice and G. E. Means, J. Biol. Chem., 246, 831 (1971).
50. M. Ottesen and B. Svensson, Compt. Rend. Trav. Lab. Carlsberg, 38, 445 (1971).
51. A. M. Crestfield, S. Moore, and W. H. Stein, J. Biol. Chem., 238, 622 (1963).
52. T. D. C. Grace, Nature, 195, 788 (1962).
53. R. D. Palmiter, Biochemistry 13, 3606 (1974).
54. G. Brawerman, in Methods in Cell Biology, D. M. Prescott, ed. Vol. VII, Academic Press, New York, 1973, p. 1.
55. R. P. Perry, J. LaTorre, D. E. Kelley, and J. R. Greenberg, Biochim. Biophys. Acta, 262, 220 (1972).
56. H. Aviv and P. Leder, Proc. Natl. Acad. Sci. USA 69, 1408 (1972).
57. J. Gielen, H. Aviv, and P. Leder, Arch. Biochem. Biophys., 163, 146 (1974).
58. R. A. Firtel and H. F. Lodish, J. Mol. Biol., 79, 295 (1973).
59. P. A. Kitos, G. Saxon, and H. Amos, Biochem. Biophys. Res. Commun., 47, 1426 (1972).
60. P. A. Kitos and H. Amos, Biochemistry, 12, 5086 (1973).
61. P. A. Kitos, T. Fuller, G. S. King, and R. T. Hersh, Biochim. Biophys. Acta, 353, 362 (1974).

62. J. DeLarco and G. Guroff, Biochem. Biophys. Res. Commun., 50, 486 (1973).

63. N. Sullivan and W. K. Roberts, Biochemistry, 12, 2395 (1973).

64. G. Brawerman, J. Mendecki, and S. Y. Lee, Biochemistry, 11, 637 (1972).

65. R. Sheldon, C. Jurale, and J. Kates, Proc. Natl. Acad. Sci. USA 69, 417 (1972).

66. R. J. Greenberg and R. P. Perry, J. Mol. Biol., 72, 3 (1972).

67. P. T. Gilham, J. Amer. Chem. Soc., 86, 4982 (1964).

68/ J. Kates, in Methods in Cell Biology, D. M. Prescott, ed. Vol. VII, Academic Press, New York, 1973, p. 531.

69. M. E. Haines, N. H. Carey, and R. D. Palmiter, Eur. J. Biochem., 43, 549 (1974).

70. R. H. Singer and S. Penman, J. Mol. Biol., 78, 321 (1973).

71. L. F. Johnson, H. T. Abelson, H. Green, and S. Penman, Cell, 1, 45 (1974).

72. R. DeWachter and W. Fiers, in Methods in Enzymology, S. P. Colowick and N. O. Kaplan, eds. Vol. XXI, Academic Press, New York, 1971, p. 167.

73. A. C. Peacock and C. W. Dingman, Biochemistry, 6, 1818 (1967).

74. T. Maniatis and M. Ptashne, Proc. Natl. Acad. Sci. USA 70, 1531 (1973).

75. L. Reijnders, P. Sloof, J. Sivae, and P. Borst, Biochim. Biophys. Acta, 324, 320 (1973).

76. R. W. Floyd, M. P. Stone, and W. K. Joklik, Anal. Biochem., 59, 599 (1974).

77. D. Z. Staynov, J. C. Pinder, and W. B. Gratzer, Biochemistry, 13, 5373 (1974).

78. H. J. Gould and P. H. Hamlyn, FEBS Letters, 30, 301 (1973).

79. J. J. Shearman, P. H. Hamlyn, and H. J. Gould, FEBS Letters, 47, 171 (1974).

80. J. C. Pinder, D. Z. Staynov, and W. B. Gratzer, Biochemistry, 13, 5373 (1974).

81. H. H. Kazazian, P. G. Snyder, and T. Cheng, Biochem. Biophys. Res. Commun., 45, 184 (1971)

82. J. Horst, J. Content, S. Mandeles, H. Fraenkel-Conrat, and P. Duesberg, J. Mol. Biol., 69, 209 (1972).

83. Y. P. L. Young and R. J. Young, Anal. Biochem., 58. 286(1974).

84. W. Gilbert and A. Maxam, Proc. Natl. Acad. Sci. USA, 70, 3581 (1973).

85. A. F. Wagner, R. L. Bugianesi, and T. Y. Shen, Biochem. Biophys. Res. Commun., 45, 184 (1971).

86. M. S. Poonian, A. J. Schlabach, and A. Weissbach, Biochemistry, 11, 533 (1972).

87. M. H. Schreier and T. Staehelin, J. Mol. Biol., 73, 329 (1973).

88. M. B. Mathews, in Essays in Biochemistry, P. N. Campbell and F. Dickay, eds., Vol. 9, Academic Press, New York 1973, p. 59.

89. B. E. Roberts and B. M. Paterson, Proc. Natl. Acad. Sci. USA 70,2330 (1973).

90. D. Efron and A. Marcus, FEBS Letters, 33, 23 (1973).

91. B. T. Schmeckpecker, S. Cory, and J. M. Adams, Mol. Biol. Reports, 1, 355 (1974).

92. K. Marcu and B. Dudock, Nucleic Acids Res., 1, 1385 (1975).

93. K. Benveniste, T. Wilczek, and R. Stern, Fed. Proc., 33, 1541 (1974).

94. H. Boedtker, R. B. Crkvenjakov, J. Last, and P. Doty, Proc. Natl. Acad. Sci. USA, 71, 4208 (1974).

95. B. M. Paterson, B. E. Roberts, and D. Yaffe, Proc. Natl. Acad. Sci. USA, 71, 4467(1974).

96. D. S. Shih, and P. Kaesberg, Proc. Natl. Acad. Sci. USA, 70, 1799 (1973).

97. J. W. Davies and P. Kaesberg, J. Gen. Virol., 25, 11 (1974).

98. C. L. Prives, H. Aviv, B. Paterson, B. E. Roberts, R. Shmuel, M. Revel, and E. Winocour, Proc. Natl. Acad. Sci. USA, 71, 302(1974).

99. J. W. Davies and P. Kaesberg, J. Vir., 12, 1434(1973).

100. A. Marcus, D. Efron, and D. Weeks, in Methods in Enzymology, K. Moldave and L. Grossman eds., Vol. XXX (Part F), Academic Press, New York, 1974, p. 749.

101. M. Green and G. F. Gerard, in Progress in Nucleic Acid Research and Molecular Biology, W. E. Cohn, ed. 14, Academic Press, New York, 1974, p. 749.

102. M. J. Modak, S. L. Marcus, and L. F. Cavalieri, Biochem. Biophys. Res. Commun., 55, 1(1973).

103. L. A. Loeb, K. D. Tartof, and E. C. Travaglini, Nature(New Biol.), 242, 66(1973).

104. J. D. Karkas, Proc. Natl. Acad. Sci. USA, 70, 3834 (1974).

105. S. C. Gulati, D. L. Kacian, and S. Spiegelman, Proc. Natl. Acad. Sci. USA, 71, 1035 (1974).

106. S. Mizutani, C. Kang, and H. M. Temin, in Methods in Enzymology, L. Grossman and K. Moldave, eds., Vol. XXIX(Part E), Academic Press, New York, 1974, p. 119.

107. I. M. Verma and D. Baltimore, ibid., p. 125.

108. J. P. Leis and J. Hurwitz, ibid., p. 143.

109. D. L. Kacian and S. Spiegelman, ibid., p. 150.

110. D. L. Kacian, K. F. Watson, A. Burny and S. Spiegelman, Biochim. Biophys. Acta, 246, 365(1971).

111. J. P. Leis and J. Hurwitz, J. Virol., 9, 130 (1972).

112. R. Ruprecht, N. C. Goodman, and S. Spiegelman, Biochim. Biophys. Acta, 294, 192 (1973).

113. I. M. Verma, G. F. Temple, H. Fan, and D. Baltimore, Nature(New Biol.), 235, 163 (1972).

114. D. L. Kacian, S. Spiegelman, A. Bank, M. Terada,

S. Metafora, L. Dow, and P. A. Marks, Nature(New Biol., 235, 163(1972).

115. J. Ross, H. Aviv, E. Scolnick, and P. Leder, Proc. Natl. Acad. Sci. USA, 69, 264(1972).

116. A. J. M. Berns, H. Bloemendal, S. J. Kaufman, and I. M. Verma, Biochem. Biophys. Res. Commun., 52, 1013 (1973).

117. A. Efstratiadis, T. Maniatis, F. C. Kafatos, A. Jeffrey, and J. Vournakis, Cell, in press.

118. P. R. Harrison, G.D. Birnie, A. Hell, S. Humphries, B. D. Young, and J. Paul, J. Mol. Biol., 84, 539 (1974).

119. T. Imaizumi, H. Diggelmann, and K. Scherrer, Proc. Natl. Acad. Sci. USA, 70, 1122 (L973).

120. J. O. Bishop and M. Rosbash, Nature (New Biol.), 241, 204 (1973).

121. L. L. Kiselev, L. Y. Frolova, K. G. Gazaryan, V. Z. Tarantul, and V. A. Engelgardt, Doklo Akad. Nauk SSSR, 213, 1203(1973).

122. J. H. Chen, G. C. Lavers, and A. Spector, Biochem. Biophys. Res. Commun., 52, 767 (1973).

123. S. E. Harris, A. R. Means, W. M. Mitchell, and B. W. O'Malley, Proc. Natl. Acad. Sci. USA, 70, 3776 (1973).

124. R. F. Cox, M. E. Haines, and J. S. Entage, Eur. J. Biochem., 49, 225 (1974).

125. H. Diggelmann, C. H. Faust, and B. Mach, Proc. Natl.

Acad. Sci. USA, 70, 693 (1973).

126. P. M. Lizardi and D. D. Brown, Cold Spring Harbor Symp. Quant. Biol., 38, 701 (1974).

127. D. P. Grandgenett, G. F. Gerard, and M. Green, Proc. Natl. Acad. Sci. USA, 70, 230 (1974).

128. S. L. Marcus, M. J. Modak, and L. F. Cavalieri, J. Virol. 14, 853 (1974).

129. A. Gaparin, J. P. McDonnell, W. Levinson, N. Quintrell, L. Fanshier, and J. M. Bishop, J. Virol., 6, 589 (1970).

130. M. Green, M. Rokutanda, K. Fujinaga, R. K. Ray, H. Rokutanda, and C. Gurgo, Proc. Natl. Acad. Sci. USA, 67, 385 (1970).

131. A. Gaparin, L. Fanshier, J. Leong, J. Jackson, W. Levinson, and J. M. Bishop, J. Virol., 7, 227 (1971).

132. C. H. Faust, H. Diggelmann, and B. Mach, Biochemistry, 12, 925 (1973).

133. H. Noll, Nature, 215, 360 (1967).

134. D. S. Auld, H. Kawaguchi, D. M. Livingston, and B. L. Vallee, Proc. Natl. Acad. Sci. USA, 71, 2091 (1974).

135. A. J. Faras, J. M. Taylor, J. P. McDonnell, W. E. Levinson, and J. M. Bishop, Biochemistry, 11, 2334 (1972).

136. P. R. Harrison, A. Hell, and J. Paul, FEBS Letters, 24, 73(1972).

137. R. J. Britten, D. E. Graham, and B. R. Neufeld, in Methods in Enzymology, Vol. XXXIX(Part E), Academic Press, New York, 1974, p. 363.

138. T. Ando, Biochim. Biophys. Acta, 114, 158 (1966).

139. W. D. Sutton, Biochim. Biophys. Acta, 240, 522 (1971).

140. W. M. Vogy, Eur. J. Biochem., 33, 192 (1973).

141. J. H. Crosa, D. J. Brenner, and S. Falkow, J. Bacteriol., 115, 904 (1973).

142. J. Leong, A. Gaparin, N. Jackson, L. Fanshier, W. Levinson, and J. M. Bishop, J. Virol., 9, 891 (1972).

143. P. Beard, J. F. Morrow, and P. Berg, J. Virol.,9, 891 (1972).

144. I. M. Verma, R. A. Firtel, H. F. Lodish, and D. Baltimore, Biochemistry, 13, 3917 (1974).

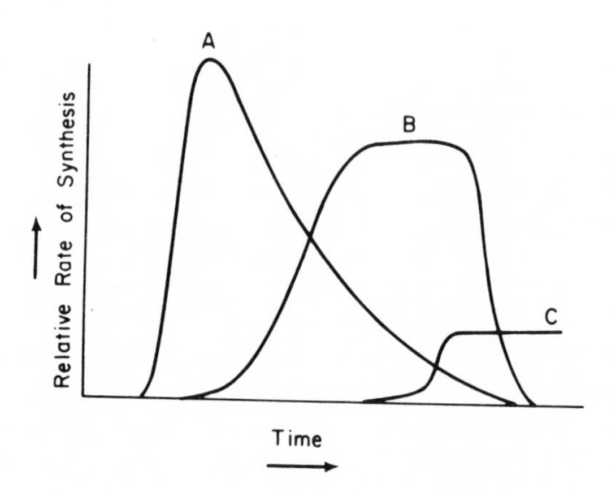


FIG. 1. Differentiation as a program of differential gene expression at the level of protein synthesis. A hypothetical example is shown, in which three differentiation-specific proteins (A, B, C) are regulated with separate developmental kinetics of synthesis. Differentiation at any time point would be characterized by a unique pattern of relative synthetic rates for these proteins (cf. Fig. 5).

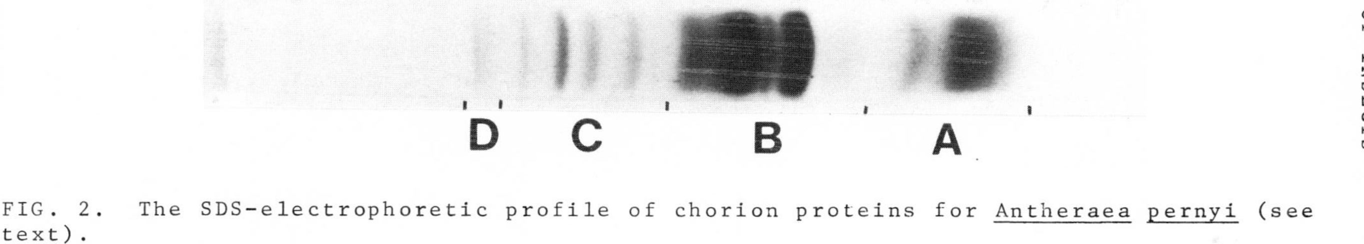


FIG. 2. The SDS-electrophoretic profile of chorion proteins for *Antheraea pernyi* (see text).

FIG. 3. Isoelectric focusing of chorion proteins from *Bombyx mori* (strain 701) on a polyacrylamide slab gel (pH 4-6 Ampholines; acidic end on the right) (see text and Ref. 1).

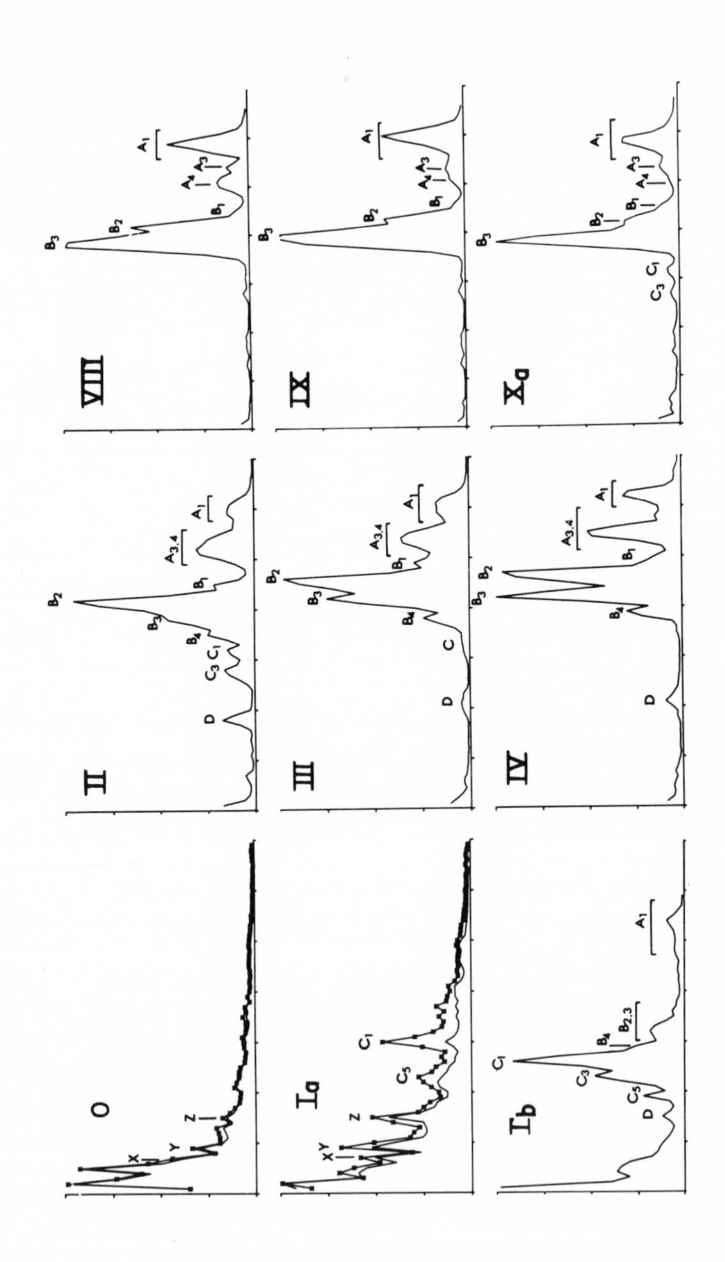

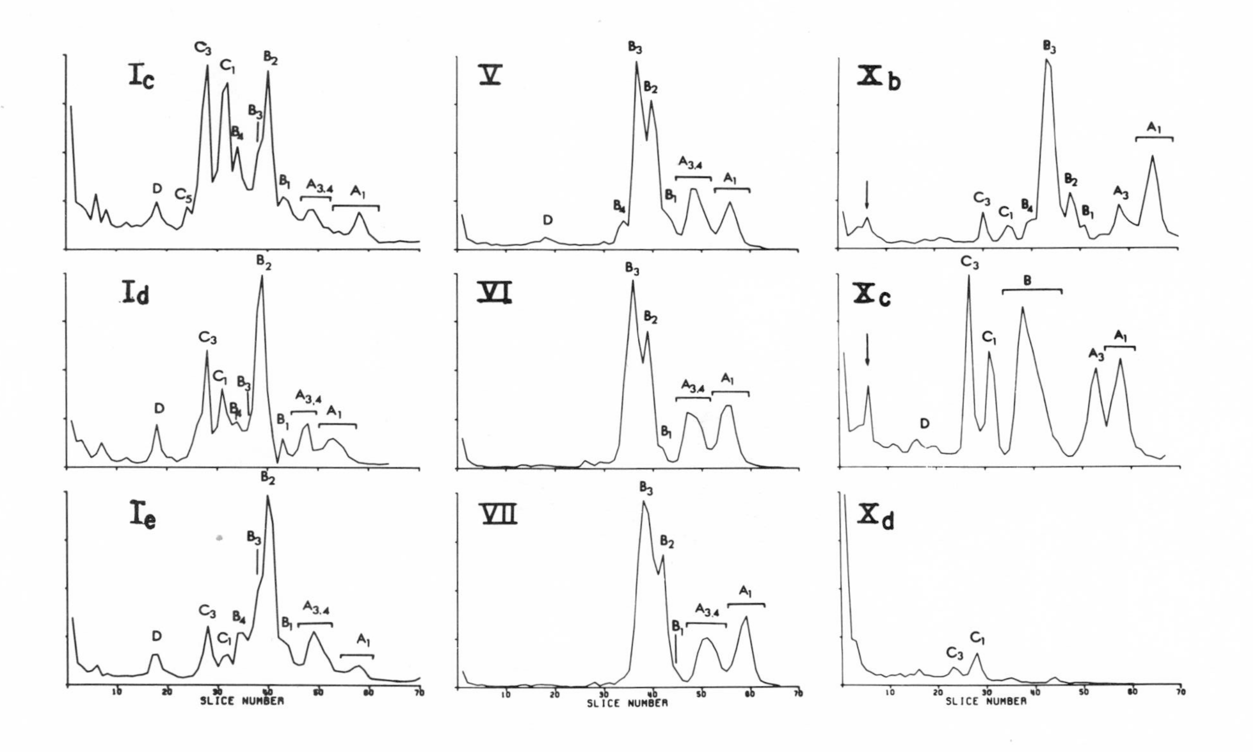


FIG. 4. Stages of choriogenesis in A. polyphemus. The relative rates of synthesis (ordinate) of proteins fractionated by mobility in an SDS-polyacrylamide gel (abscissa) are shown for defined developmental stages (individual panels). The relative rates are normalized for each sample and are not comparable in absolute terms between panels. From Kafatos et al. [1] and Paul and Kafatos [6].

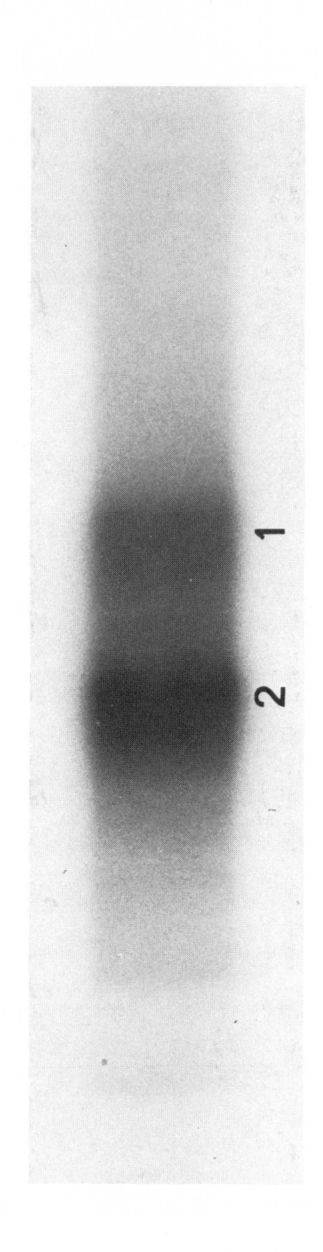


FIG. 5. Autoradiogram of $^{32}P$-labeled chorion mRNA from *A. polyphemus*. In vivo-labeled RNA was extracted from polysomes, fractionated on a sucrose density gradient, purified by binding to a nitrocellulose filter, and analyzed on a polyacrylamide slab gel (6% monomer, 7M urea). (From Vournakis et al. [10].)

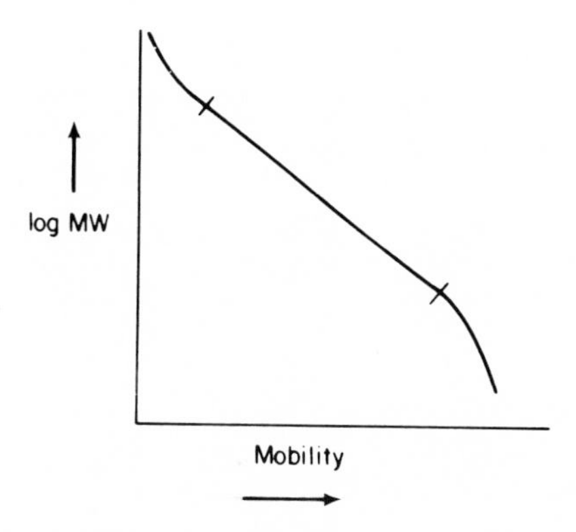


FIG. 6. The usual relationship between protein size and mobility on SDS-polyacrylamide gels. The marks outline the useful range for protein fractionation (see text and Refs. 22 and 23).

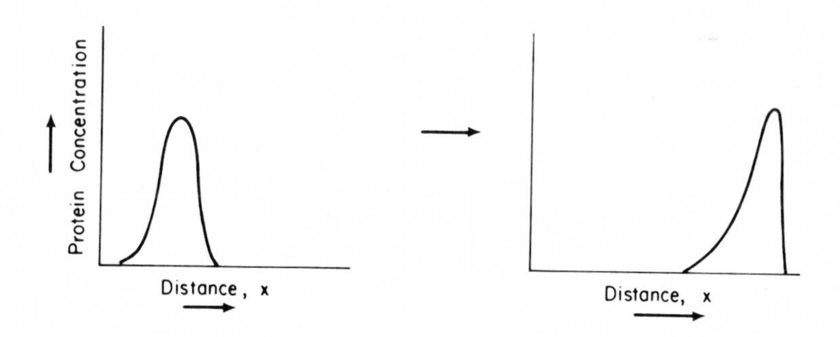


FIG. 7. Distortion of a band during electrophoresis, resulting from overloading. An initially symmetrical distribution (left) becomes asymmetric (right), as the protein migrates faster than elsewhere when in the highest concentration regions of the band.

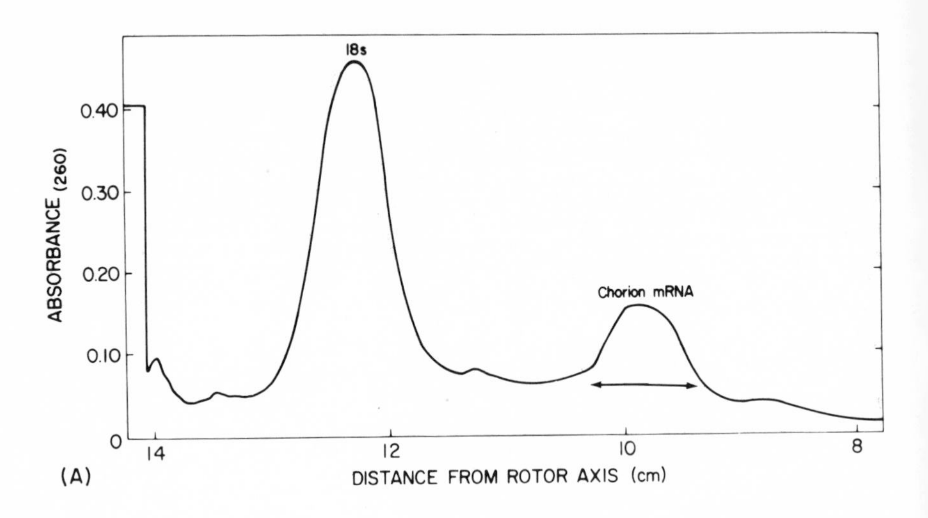


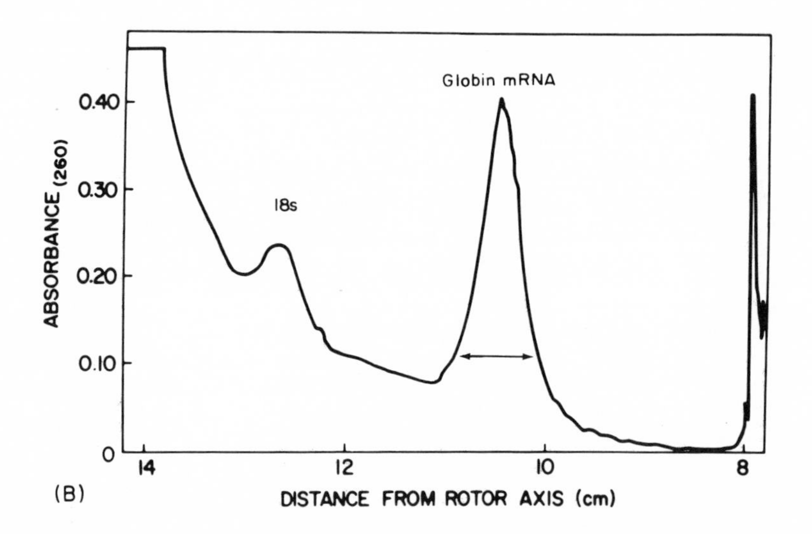


FIG. 8. The bound material from the first oligo(dT)-cellulose column (Table 3) was centrifuged through a 5-20% sucrose gradient as described in the text. (a) chorion RNA, (b) globin RNA. Twelve 1-ml fractions were collected, while the absorbance was monitored at 260 nm. The indicated fractions were pooled, EtOH-precipitated, and further purified through a second oligo(dT)-cellulose column (Table 3). Methylene Blue stain.

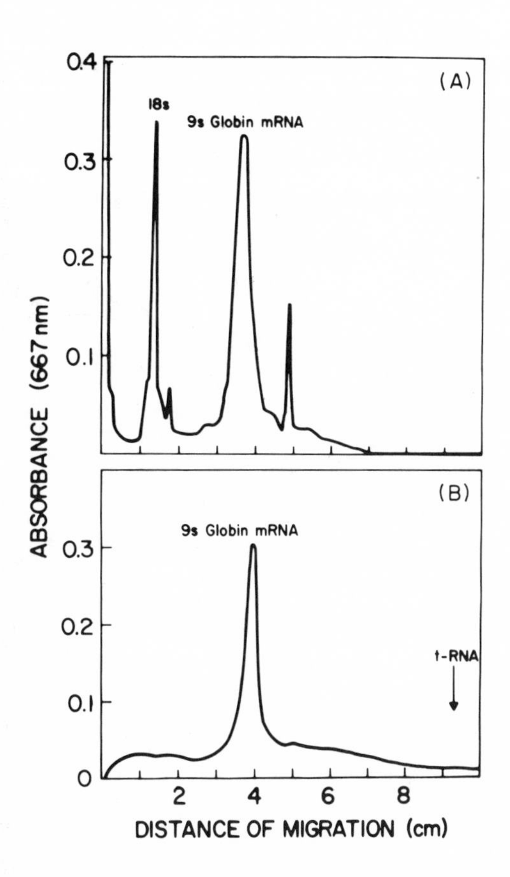


FIG. 9. Electrophoresis in cylindrical urea-acrylamide gels (see Sec. VD) of aliquots (a) of the RNA pooled from the sucrose gradient (Fig. 8B), (b) of the RNA shown in Fig. 9A, after binding to a second oligo(dT)-cellulose column (Table 3). Methylene Blue stain.

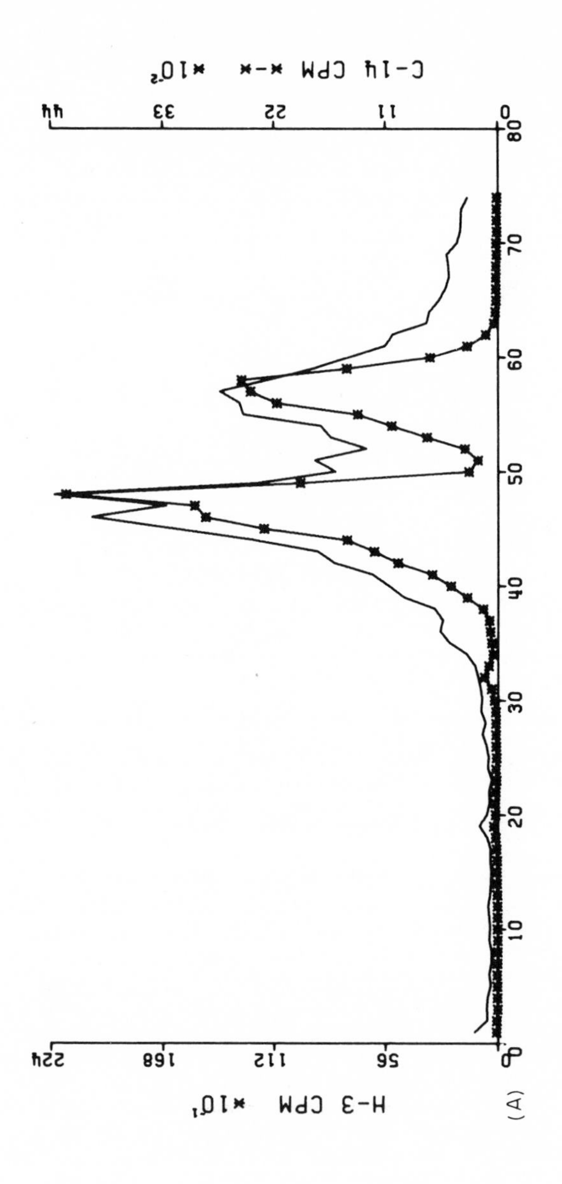
C-14 CPM *-* *10²
0
11
22
33
44
H-3 CPM *10¹
0
56
112
168
224
0
10
20
30
40
50
60
70
80
(A)

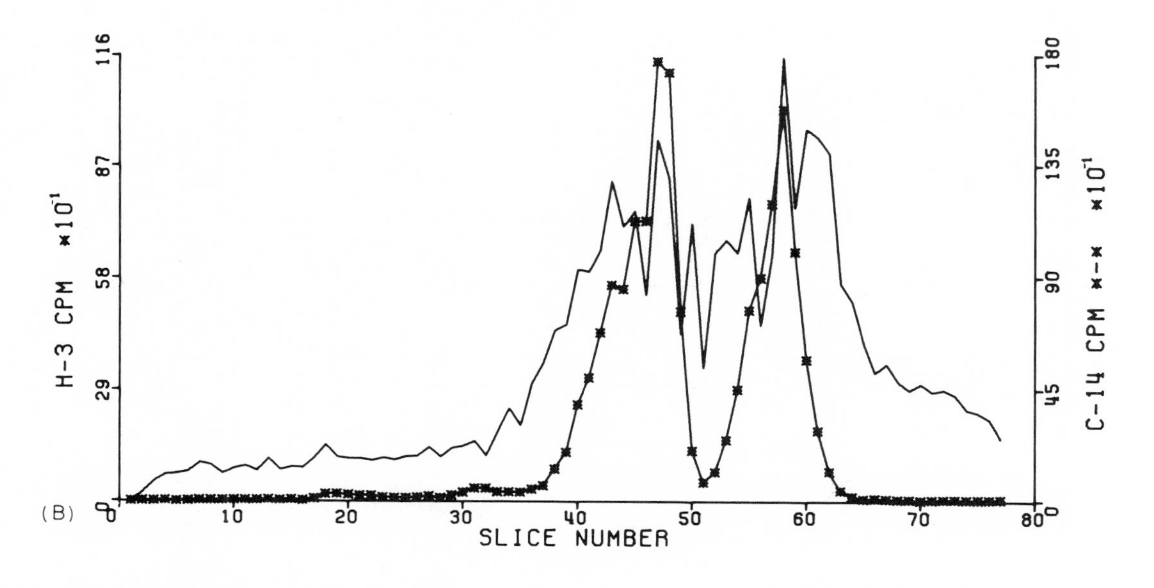


FIG. 10. Cell-free translation in the wheat germ extract, of partially purified chorion mRNA from A. pernyi. Aliquots of [$^3$H]-leucine-labeled material synthesized in the presence of 120mM $K^+$ (A) or 100mM $K^+$ (optimum for incorporation) (B), were dissolved and analyzed on SDS-polyacrylamide gels. Authentic $^{14}$C -labeled chorion is added as an internal standard prior to electrophoresis (see text).

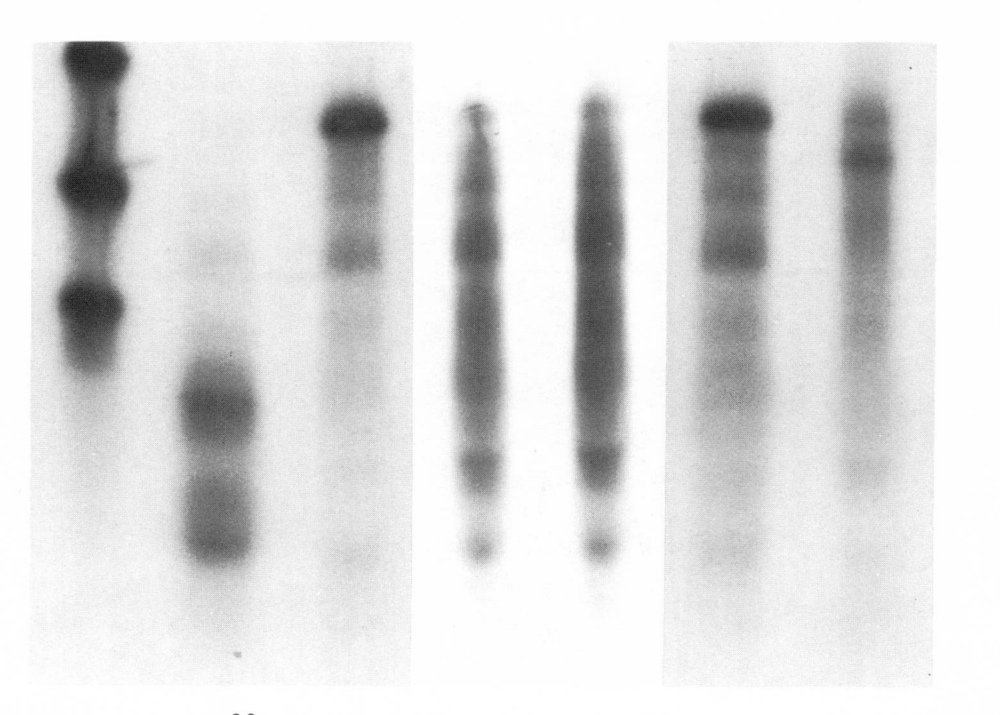

FIG. 11. Autoradiogram of [$^{32}$P]-globin cDNA electrophoresed on a 5% polyacrylamide slab gel in 98% formamide, after gel filtration through Sephadex G-100. The basic reaction mixture was as in Table 5, column 4. The label was introduced by [$^{32}$P]-dGTP. Slot 1:[$^{32}$P]-λDNA markers (900, 440, and 230 nucleotides long, Hin digests), Slot 2: Products in the presence of dNTPs at a concentration of 5 μM each, Slot 3: dNTPs at 100 μM, each, Slot 4: Labeled dNTPs at 5 μM, unlabeled at 100 μM, Slot 5: Labeled dNTPs at 5 μM, unlabeled at 500 μM, Slot 6: Labeled dNTPs at 25 μM, unlabeled at 100 μM, Slot 7: Reverse transcription products of chorion mRNAs (30 μg/ml) in the presence of dNTPs, at a concentration of 100 μM each.

# Chapter 2

# ANALYSIS OF DATA FROM DNA-DNA AND DNA-RNA HYBRIDIZATION EXPERIMENTS

Patrick J. Williamson
National Institutes of Health
Bethesda, Maryland

# I. INTRODUCTION

## A. The Requirement for an Assay

Eukaryotic organism mRNA has been studied since at least 1965 [1,2], when the study of the messenger populations of cultured cells began in earnest. The study of heterogenous populations of molecules can be frustrating, however, and it was not uncommon during this period to see and hear in discussions that most of the answers would just fall out if only "we could isolate and study a single messenger." The possession of a tube containing a pure RNA species and the intent to learn its functional characterstics is analogous to the possession of a pure protein, the function of which is unknown. Many studies of mRNA require, not the ability to study the isolated purified RNA molecule, but the ability to recognize the mRNA and measure its concentration in the unpurified state. This type of measurement requires not the molecule itself, but rather an assay for that molecule. If no assay for the mRNA is available, a complete physical characterization of the molecule, including its primary sequence, may prove misleading or totally unrelated to the functional properties of the molecule.

## B. The Translation Assay

To characterize mRNAs, cell-free systems capable of translating an mRNA into the protein it codes for are generally used since messenger activity is the fundamental property of a presumptive mRNA. Under appropriate conditions, the amount of a specific protein synthesized provides a measurement of the amount of mRNA added to such a cell-free system.

Although useful, the translation assay suffers from several disadvantages. The first problem is that the assay system itself is enormously complex, ranging from mixtures of purified components supplemented with partially characterized "wash factors" to unfractionated extracts containing huge quantities of extraneous proteins, and including unknown quantities of nucleases and proteases. Another problem with translation assays is that the amount of purified messenger required to produce measureable stimulation of synthesis of a specific protein is a fraction of a microgram: even when the use of antibodies or extracts with low background allcws increased sensitivity, 10-100 ng of messenger remains the lower range of detectability. Finally, in many experiments, such as those involving heterogeneous nuclear RNA, the ability or efficiency of messenger sequences to function in protein synthesis is one of the questions at issue and, hence, is not a satisfactory property on which to base an assay.

## C. The Hybridization Assay

These drawbacks to the translation assay emphasize the need for an mRNA assay that is simple, sensitive, and functionally neutral. Mother Nature, working in collaboration with Temin [3] and Baltimore [4], has provided the key to such an assay in an enzyme from RNA tumor viruses that copies RNA into DNA. The discovery that, under the appropriate conditions, this RNA-dependent DNA polymerase or "reverse transcriptase" would copy mRNA into a complementary single-stranded DNA [5-7] opened the door to a hybridization assay with precisely the desired properties of simplicity, sensitivity, and neutrality. When isotopically labeled to high specific activities, this complementary DNA (cDNA) can be used in a hybridization assay with precisely the desired properties of simplicity, sensitivity, and neutrality. When isotopically labeled to high specific activities, this complementary DNA (cDNA) can be used in a hybridization assay where the appearance of the label in hybrid form (i.e., in a double helix) would imply the presence of its complement, the mRNA sequence. From the rate of hybrid formation and a knowledge of the rate constant of hybridization, the concentration of RNA can be quantitated. Hybridization reactions require only that the cDNA probe and the RNA be incubated at an appropriate temperature under suitable conditions of ionic strength and pH. Controls for such variables as nonspecific background and adsorption, precipitation, or degradation of the probe or RNA are simple to run and can be included in parallel in almost every experimental situation. Moreover, the DNA, once isolated and purified, can be expected to remain reasonably stable and reliable with only minimum precautions, since it is not required that it maintain any complex structure or biologic activity. Finally, the reactants and reaction are sufficiently defined to permit quantitative evaluation and theoretical interpretation of data with a minimum of assumptions about unknowable variables.

The sensitivity of a hybridization assay using a labeled single-stranded DNA probe is limited by only two considerations: (a) the specific activity to which the probe can be labeled and (b) the minimum reactant concentration commensurate with practical reaction kinetics and yields. If we take the limit of commercially available isotope specific activities to be about 25 Ci/mmole for [$^{3}$H] nucleoside triphosphates, and assume the use of all four labeled DNTPs in preparing the probe, the upper limit for DNA specific activities is roughly $10^{8}$ cpm/μg. If a statistically believable number of counts per minute in hybrid is about 100, the specific activities available are sufficient to permit the detection of as little as 1 pg($10^{-3}$ng) of RNA sequences. The detection of this amount of RNA, however, requires that the nucleic acid concentrations be high enough to allow hybridization to occur at reasonable rates, where the primary limiting factor is the minimum volume of incubation that is technically tolerable. In a reaction volume of 2 μl, the hybridization of 10 pg of RNA and 50 pg of DNA is essentially complete in less than 48 hours, a time comparable to that used for high $C_o t$ DNA renaturations (see below) and RNA-DNA hybridizations. Furthermore, 60% of the RNA or more can be brought into hybrid form at equilibrium under these conditions [6]. Although 2 μl is probably not far from the lower volume limit for convenient work, it is a sufficiently small volume to allow assays within an order of magnitude of the limit of sensitivity imposed by isotope considerations.

Finally, the hybridization test for messenger sequences does not tell us very much about the functional state of the RNA in question. Most hybridizations are performed at 25°C below the melting temperature of double helices, conditions under which most of the secondary structure in mRNA molecules is melted away [9]. Moreover, the general practice of heating hybridization samples before incubation to well above the melting temperature of double helices ensures that even messenger sequences totally sequestered into long double-helical regions are made available to the probe. In the case where these helices are present in high concentrations, the cDNA will act as a tracer for their renaturation; only if the messenger sequences are in "palindrome" structures will they be likely to go undetected; indeed, it is difficult to imagine a method that could detect palindromes under these conditions.

## II. METHODOLOGY

### A. Synthesis of DNA

The methods used for the hybridization assay are straightforward, although they vary in detail from one group to another [5-8]. In general, purified mRNA is incubated with purified reverse transcriptase from avian myeloblastosis virus in a medium containing buffer, salt, magnesium, dithiothreitol, the four dNTPs, of which one or more is generally labeled, and a molar excess of a primer of oligo(dT) about 12-18 residues long. Actinomycin D is also included to ensure that the product is single stranded. After incubation for various periods of time, the template is degraded with alkali, and the cDNA is separated from small molecules by Sephadex chromatography. The yield of DNA with respect to input RNA in these reactions has been claimed to be as high as 80%, although, in this laboratory, yields of 1-10% have been the rule.

A 0.1 to 0.3 ml reaction mixture [55] contains 60 mM NaCl, 6 mM $MgAc_2$, 10 mM dithiothreitol, 50 mM Tris-HCl (pH 8.3), 300 μM each of three unlabeled deoxyribonucleotide triphosphates, one labeled deoxyribonucleotide triphosphate (usually [$^3$H] CTP), 20 μg/ml of actinomycin D, 5 μg/ml of oligo$(dT)_{12-18}$ (Collaborative Research, Inc., Waltham, Mass.), 10 μg/ml of globin mRNA, and 15% (by volume) of RNA-dependent DNA polymerase (as eluted from phosphocellulose). After hydrolysis of the template for

15 minutes at $100^{o}$ C in 0.33M NaOH, HCl is added to neutralize the NaOH. The solution is then made 0.05M in $NH_4HCO_3$, from which divalent metal ions were removed by passage through a Chelex (BioRad) column. The cDNA is lyophilized, then dissolved in 2 mM $Na_2EDTA$ (pH 7.0) and stored at $-20^{0}$ C.

Although this synthetic reaction is remarkably free of technical idiosyncrasies,a few details deserve mention. Low concentrations of added dNTPs result in decreased yield and product size. In general, the nonradioactive substrates are included at concentrations of 250-1000 μM and do not constitute a problem. But, economic considerations often preclude the use of such concentrations of a radioactive dNTP, especially when the probe is prepared with a particularly high specific activity. Nevertheless, caution should be taken to maintain the concentration as high as possible, and at least as high as 25 μM to prevent inhibition of the reaction. Moreover, some care is advised in the choice of a labeled dNTP. The use of radioactive dTTP has the disadvantage that any synthesis of DNA on the poly(A) sequence, prior to synthesis on the heterogeneous mRNA sequences, will be preferentially marked by labeled thymidine. In addition, this labeled region will almost always be scored as single stranded, because of the low melting temperature of the A-T helix, so that the final extent of hybridization will be artificially low. $[^{3}H]$dGTP should also be avoided.

because of its tendency to lose its label. The loss of label probably occurs from exchange, during extended incubations at the high temperatures of aqueous hybridizations.

B. Hybridization-Technology

The hybridization assay involves simply incubating together the DNA probe and the RNA sample under the appropriate ionic condtions. Normally, the volume of these incubations is kept at a minimum, resulting in a requirement to prevent evaporation. A common solution is to carry out the incubations in droplets sealed inside glass capillaries; the use of droplets under oil is another possibility. The choice of ionic conditions is a matter of judgment; however, either 0.3M NaCl-0.03M Na citrate (2 x SSC) or 0.2M sodium phosphate (pH 6.8) are popular choices. For DNA-DNA renaturations (see below), several ionic conditions are often used in the interest of reducing the extremely long incubation times. An ionic strength of 0.3M ($Na^+$) corresponds to a hybridization temperature (chosen about 25° C below the helix melting temperature) of 68° C.

An important problem in these, as in any hybridizations involving RNA, is the extreme sensitivity of the RNA to degradation during prolonged incubation. Degradation generally can be minimized by one of several approaches.

But RNA is particularly susceptible to hydrolytic degradation catalyzed by divalent cations; such a degradation can be routinely avoided by treatment of buffers with ion-exchange resins (e.g., Chelex), to remove divalent cations, and by the inclusion of small amounts of EDTA in the hybridization medium to scavenge the remaining divalent ions. A more difficult problem is the susceptibility of RNA to enzymatic degradation by the ubiquitous RNases that plague all work with RNA. Although exclusion of RNase by suitable precautions taken during the preparation of the reactants and buffers is necessary, probably the most effective method of countering trace RNase activity is the inclusion of 0.2-0.5% SDS in the hybridization buffer to denature any RNase molecules that survive the purification procedures. Finally, the rate of both of these processes, and of thermal degradation as well, can be slowed by reducing the temperature of hybridization -- either by decreasing the ionic strength or even more dramatically by including formamide, a denaturant, in the hybridization buffer. Inclusion of formamide at a final concentration of 50% by volume at the salt concentrations listed above results in sufficient helix destabilization to reduce the maximal hybridizing temperature to roughly 37° C, a convenient value. Unfortunately, the lower temperature and higher solvent viscosity of the aqueous conditions reduce the hybridization rate by a factor of five to six from the rate in aqueous media, so that except for very long reactions, much of the increased RNA stability is offset by the longer incubation times required to obtain the same net result obtained in purely aqueous reactions.

## C. Hybridization Assay*

The detection of helix formation is the heart of the hybridization assay. Because of the small quantities of probe used, optical methods for the determination of the extent of hybridization are impractical; only two methods present themselves for use with such small amounts of nucleic acid. The first and more classical method involves chromatography on hydroxyapatite to separate double-stranded from single-stranded material. The hydroxyapatite assay can be used without recourse to nuclease to trim single-stranded tails; therefore, the hybrids can be recovered in substantially intact form for further experimentation. As yet, however, this advantage has not been widely exploited, and the laboriousness of the requirement to run a column for every experimental point has led to widespread adoption of the second method for determining the amount of cDNA in hybrid, the use of nucleases for single-stranded DNA.

---

*See also Chap. 1.

A single-strand specific nuclease is particularly convenient, because the amount of hybrid formed is simply the amount of cDNA remaining large (TCA-precipitable) after incubation of a sample with the nuclease. Extensive use has been made of the single-strand nuclease from *Aspergillus oryzae*, for which extremely simple isolation procedures are available [10,11]; they yield stable enzyme preparations of high specific activity from cheap sources, advantages which have resulted in the commercial availability of the enzyme (Miles). This enzyme is not without its disadvantages, however. The nuclease behaves abnormally when the DNA substrate is present in very low initial concentrations [10]. As a result, carrier single-stranded DNA is generally included in nuclease assays to maintain uniform initial substrate concentrations. Also, the nuclease activity is inhibited (probably because the enzyme requires zinc, alhhough this explanation is not entirely satisfactory) by EDTA, citrate, and phosphate [11], all anions common to the hybridization solutions normally used. The inhibitory effects of these anions is normally avioded by extreme dilution, 100-fold or more, into the nuclease buffers, a dilution aided by the small volumes normally used in these incubations. Controls incubated without complementary RNA should normally be included to monitor for unexpected inhibitions of this sort.

Hybridizations are performed in 2 μl reaction mixtures contained in graduated 5 μl Clay-Adams Micropets [8]. Reaction mixtures contain 0.2M $NaH_2PO_4$ (pH 6.8), 0.5% SDS, 1 mM EDTA, 200 μg/ml oligo(U) (Calbiochem), about 0.05 μg/ml $[^3H]$cDNA (2-10 x $10^6$ cpm/μg), and as little as 5 ng/ml, but more usually about 5 μg/ml, of globin mRNA. When long hybridization times are anticipated, the reaction mixture sometimes contains 50% (by volume) redistilled formamide, and the hybridizations are performed at a correspondingly lower temperature. Hybridization mixtures are denatured in boiling water for three to five minutes and are incubated at $68^o$ C for aqueous hybridizations and at $37^o$ C for hybridizations in 50% formamide. After incubation for appropriate times, the hybridization mixtures are flushed into 0.5 ml of a buffer containing 10 mM NaCl, 0.03 mM $ZnSO_4$, 25 μg/ml of denatured calf thymus DNA, 30 mM NaAc (pH 4.5), and 1% (by volume)S1 nuclease (5400 units/ml in 25% glycerol). This mixture is incubated for 45 minutes at $45^o$ C, then nuclease-resistant cDNA is determined after precipitation with 5% TCA.

### D. Hybridization Data Analysis

Determination of the concentration of a specific mRNA sequence in an unknown RNA sample using the cDNA probe can involve either the extent or the rate of the reaction. In the first case, the number of complementary sequences is determined in incubations in which, at equilibrium, the amount of hybrid formed is below saturation of the DNA and, hence, proportional to the concentration of RNA sequences. Under the simplest conditions, the DNA would be present in excess, driving the reaction to completion in a predictable time no matter what the RNA concentration. The amount of hybrid formed, determined either from the amount of DNA or of RNA brought into the duplex, would directly measure the amount of RNA species originally present. The difficulty with this method is that the amount of available cDNA is small, so that reactions must be slow and incomplete, and they often require preliminary experiments to determine the range of RNA concentrations required to bring the reaction below completion, experiments which themselves have been raised to the level of an assay [12]. In the absence of a great cDNA excess, the alternative is work at low and approximately equivalent DNA and RNA concentrations, using a calibration curve to convert the amount of hybrid formed into an RNA concentration. This method is especially convenient for experiments in which many fractions, for example in a sucrose gradient, are tested for their mRNA sequence content. The difficulty of the method for quantitative work lies in the curvature of the calibration curve and the laboriousness of its construction in routine experiments.

The second general approach to measurement of the specific sequence content of an RNA preparation is the determination of the rate at which it induces the formation of a cDNA-RNA hybrid. In general, the hybridization reaction can be considered to be second order, since, with the method of analysis chosen, any hybrids that have successfully nucleated have ample time to complete the "zipping" reaction required to produce the complete helix before the nuclease treatment commences. Thus, the reaction kinetics can be described by the equation

$$\frac{d(H)}{dt} = k_1(R)(D) - k_{-1}(H) \quad (1)$$

where (H), (R), and (D) are the concentrations of hybrid, RNA, and DNA, $k_1$ is the forward rate constant, and $k_{-1}$ the reverse rate constant. Solutions to this equation describing the formation of hybrid as a function of time can be obtained either exactly or with the assistance of a formidable battery of approximations. One of the most popular of these approximations is the consideration of only those reactions in which the RNA sequences are present in considerable excess over the DNA sequences, so that their concentration does not change appreciably over the couse of the reaction. Under these conditions the back reaction can usually be ignored and the above equation becomes

$$\frac{d(H)}{dt} = k'(D) \quad (2)$$

where k' equals $k_1(R_o)$. This pseudo first-order reaction then has the solution

$$H = D_o\ (1- e^{-k't}) \tag{3}$$

The DNA excess equation is symmetrically similar to this one. Under these conditions, the extent of the reaction ($H/D_o$) at any given time is not dependent on the initial DNA concentration at all, but only on the concentration of the excess RNA.

However, this condition of cRNA excess is not always easily attainable, so that in some reactions true second-order kinetics still obtain. A common approximation under these conditions is consideration of the reaction as irreversible [13], so that the $k_{-1}$ term can be ignored. This leaves the rate equation

$$\frac{d(H)}{dt} = k_1(R)(D) \tag{4}$$

which has the solution

$$\frac{(D_o - H)}{(R_o - H)} = \frac{D_o}{R_o}\ e^{(D_o - R_o)k_1 t} \quad . \tag{5}$$

Finally, it is possible to integrate the equation without any approximations at all to give a solution

$$\frac{(U^1 - H)}{(U^2 - H)} = \frac{U^1}{U^2}\ e^{(U^1 - U^2)k_1 t} \tag{6}$$

where

$$U^1 = \frac{1}{2}\ ([R_o + D_o + k-1/k1] + \ [(R_o + D_o + k-1/k1)^2 - 4R_oD_o]^{1/2}) \tag{7}$$

and

$$U^2 = \frac{1}{2}([R_o+D_o+k-1/k1] - [(R_o+D_o+k-1/k1)^2-4R_oD_o]^{1/2}) \quad (8)$$

The use of these various equations to determine kinetic constants has generally involved any one of several methods of rearranging and plotting the data obtained in actual incubations. A favorite method, derived from work on DNA renaturation [14], plots the extent of hybrid formation versus the log of the product of the initial concentration of RNA and the time ($R_ot$). The value at half-reaction ($R_ot_{1/2}$) is then taken to be simply related to the rate constant of hybridization. This method has the advantages of presenting a large amount of data on a single figure and of combining data taken at several reactant concentrations. In addition, the method often emphasizes different kinetic classes of reacting species, such as the repeated versus the unique sequences in DNA-DNA renaturations. Against these advantages are several disadvantages. As originally derived, these plots were used in renaturation reactions where the concentration of reacting species were, by definition, equal. In the RNA excess reaction, where these plots have been commonly used, the shape of the curves obtained cannot be expected to mimic the shape of the original $C_ot$ plots [13]. They are sufficiently close, however, to make any transition to second-order kinetics difficult to discern. During this period of transition, estimation of initial complementary sequence concentrations by extrapolation from reactions run at pseudo first-order conditions is necessarily inaccurate. Coupled with the requirement for computer methods of fitting curves to the experimental points on such plots, these disadvantages are enough to recommend consideration of other methods.

The pseudo first-order reaction offers at first glance a method for displaying experimental results in a plot that should yield a straight line. Rearrangement of Eq. (3) yields

$$\ell og(1-H/D_0) = k_1 R_o t \tag{9}$$

so that the log of 1-H/D plotted against time should yield a straight line with a negative slope of k' or $k_1R_o$. Although appealing, because no assumptions are required, this method has several disadvantages. For various reasons [8], the hybridization reaction does not always proceed to completion with 100% of the input cDNA counts in hybrid form. This fact implies that the relevant $D_o$ may not be known in advance, making the insertion of a value in the above equation risky. This treatment also has the disadvantage that early time points (where the pseudo first-order assumption is most applicable in marginal situations) are squeezed into a very small region of the plot by the nature of the logarithmic transformation in the region just below 1.0. At late times in the reaction, the value within the parentheses becomes a small difference between two large numbers; any error in their determination is magnified enormously by the logarithmic transformation near values of zero. Moreover, in a curve-fitting method, such as least squares, these latter values are weighted very heavily in the determination of the slope of the line, so that errors in this region are represented with a vengeance in the final result. Finally, this relationship is dependent on the RNA excess assumption and is, therefore, useless in cases where complementary RNA sequences are present in a concentration comparable to that of the DNA.

A more general method that has been verified empirically [16-19], but which is in some disrepute because no one has justified it theoretically is the use of the double-reciprocal plot of hybridization data, a plotting method that results in a straight line. Bishop has noted [15] that if the exponential term in Eq. (3) is expanded, the higher-power terms can be neglected as long as $k_1R_ot$ is less than one, and the resulting equation can be rearranged as

$$\frac{1}{H} = \frac{1}{k_1R_oD_ot} + \frac{1}{D_o} \qquad (10)$$

An extension of this method to reactions not in great RNA excess can be made in the following fashion. Equation (5) can be rearranged to give

$$\frac{\ell og(R_oD_o - R_oH)}{R_oD_o - D_oH} = (D_o - R_o)K_1t \quad . \tag{11}$$

This equation can be simplified using the expansion

$$\ell og(x) = 2\ \frac{x-1}{x+1} + \frac{1}{3}\ \frac{(x-1)^3}{x+1} + \frac{1}{5}\ \frac{(x-1)^5}{x+1} + \ . \ . \ . \tag{12}$$

When x has a value of roughly one, as it does early in the reaction described by Eq. (11), the term $(x-1)/(x+1)$ is very close to zero, and hence the higher-power terms are vanishingly small and can be ignored. When the left side of Eq. (11) is approximated in this way and the resulting expression is simplified

$$\frac{1}{H} = \frac{1}{R_oD_ok_1t} + \frac{1}{2}\left(\frac{1}{D_o} + \frac{1}{R_o}\right) \tag{13}$$

that is, a straight line is again expected in the double-reciprocal transformation, with the same slope as in the RNA excess case, but the intercept is now an average of the reciprocals of the RNA and DNA initial concentrations.

Finally, this approximation can easily be expanded to include the effects of the reverse rate constant, $k_{-1}$. If the approximation of Eq. (12) is applied to Eq. (6),

$$\frac{1}{H} = \frac{1}{R_oD_ok_1t} + \frac{1}{2}\left(\frac{1}{D_o} + \frac{1}{R_oD_ok_{eq}} + \frac{1}{R_o}\right) \tag{14}$$

where $k_{eq} = k_1/k_{-1}$. Thus, the equations predict a straight-line relationship between the reciprocals of the amount of hybrid formed and time, with a characteristic slope that does not contain a contribution from the reverse reaction. These assumptions do not alter the value of the intercept of the extrapolated line, so that the use of this number in marginal cases is risky.

The method of plotting hybridization kinetic data in double-reciprocal form has several advantages. It does not require an accurate knowledge of $D_o$ in advance, and it often provides this value as a by-product of the plot. Furthermore, the method is very insensitive to any transitions between pseudo first-order and second-order reactions, so that reactions performed in the marginal region between these reaction types can be analyzed directly. Finally, elimination of the logarithmic transformation expands the range of useful data for curve-fitting techniques. But, the method has the drawbacks that the intercept value may be deceiving and, more importantly, that the values most heavily weighted in a normal least-squares technique of curve fitting applied to these plots are early time points,the values which are most strongly affected by errors in estimates of backgrounds, etc. Thus, data should be analyzed after elimination of those values derived from less than twice the background value of counts per minute.

## E. Limitations

Use of the cDNA synthesized by reverse transcriptase has completely dominated hybridization assays, primarily because of its simplicity. Nevertheless, it has its drawbacks, which should be taken into account, especially in areas where its use has not been already subjected to careful scrutiny. The first of these relates to its lack of universality. Since synthesis is initiated from a primer of oligo(dT) bound to the poly(A) sequence at the end of the mRNA, it is not possible to extend the technique to RNA molecules that do not possess poly(A) sequences; ribosomal RNA and histone mRNA are well-known examples.

It may eventually prove feasible to circumvent this problem by addition of poly(A) sequences to RNAs with RNA ligase or by use of primers other than oligo(dT); thus, synthesis of cDNA can be primed with oligo(dG) [20, 21], although elucidation of the generality and precise significance of these observations will require further experimentation. Eventually, it may prove possible to design specific primers given a knowledge of portions of the primary sequence of the template RNA. This approach could eliminate another problem associated with the use of the oligo(dT) primer, its inability to discriminate between poly(A)-containing mRNAs, so that even the cDNA to the highly purified globin mRNA contains some contaminating species [22]. Although this problem poses few technical limitations in the case of globin mRNA, which is isolated from a highly differentiated cell, it may prove less surmountable in the case of mRNA preparations from less specialized cells, as, for example, the mRNA for immunoglobin light chain, where such difficulties have already been reported [23].

Use of the oligo(dT) primer gives rise to another problem with the cDNA probe; synthesis of cDNA begins at the 3'-OH end of the mRNA molecule, where the poly(A) region is located. Unfortunately, there is evidence that the structural gene sequence does not continue to the 3'-OH end of the molecule, at least in the case of human globin mRNA [24]; hence, the enzyme begins copying sequences that cannot be assumed to be unique to the mRNA under study. This problem would be largely irrelevant if the enzyme simply copied past this nonstructural sequence into the body of the mRNA in all cases. However, the reverse transcriptase in general synthesizes cDNA molecules that sediment in sucrose gradients at about 6S [5,6,21,25-27] , regardless of the size of the mRNA template. Although a 6S cDNA corresponds to much of the globin messenger, with larger mRNAs this is but a small fraction of the total mRNA, a fact that must be borne in mind when the results of hybridizations using these probes are assessed.

In addition to this problem of the small size of the product obtained, there may be an additional problem of low yields (a few percent of the input RNA nucleotides) of cDNA as well. It can be demonstrated that the DNA is complementary to the major component of at least the globin mRNA preparation [8], so that this low yield does not result from copying a minor, contaminating RNA species. Nevertheless, the low yield poses technical problems. Although the production of adequate amounts of highly labeled cDNA remains unaffected, certain experiments are ruled out because they require large masses of cDNA, rather than simply a large number of counts per minute. Since the production of neither mRNA nor reverse transcriptase can easily be scaled up indefinitely, experiments at cDNA excess have to date remained largely unperformable.

### F. Alternatives

The use of reverse transcriptase and its DNA product is not the only way in which the advantages of hybridization was first realized by the use of an RNA probe. In the experiments of Melli and Pemberton [28], the RNA polymerase from *Micrococcus luteus* (nee *lysodeikticus*) was used to synthesize a complement to the globin mRNA molecule, which was then used in tests for mRNA sequences in heterogenous nuclear RNA. The problem with these experiments is that the product was RNA; a double-stranded probe was therefore inescapable because the template and product could not be separated from one another. Thus, the principle of assay became the measurement of a change in the rate of renaturation of the probe with itself, an approach that deprives the method of much of its simplicity and sensitivity.

Scaling up the DNA synthetic reaction is limited in part by the difficulties in obtaining large quantities of the tumor that is the source of the enzyme. Attempts have been made to replace the reverse transcriptase with other more easily and generally available enzymes -- in particular, DNA polymerase I from E. coli. Although early attempts to use this enzyme were not very encouraging [29-31], under appropriate conditions, such as the use of high enzyme to template ratios and the use of the inhibitor distamycin to prevent (dA · dT) synthesis, the enzyme will copy mRNA faithfully to yield a useful DNA product [32-34]. When fully optimized, the DNA polymerase reaction may make the production of relatively large quantities of cDNA feasible, allowing several experiments which at the moment are technically unrealistic to be performed.

An interesting alternative to the use of the complementary sequence is available in the labeled mRNA molecule itself. In some instances, of course, the labeled mRNA is nearly as direct a tool as the cDNA, as in experiments where the labeled probe is used as a tracer in DNA-DNA renaturation experiments. Even more interesting, however, is the use of the RNA in a competition assay; such an assay has been carried out in the case of the histone mRNA [35], which lacks poly(A) and cannot therefore be copied by the reverse transcriptase. In this type of assay, the labeled RNA is hybridized to a known amount of DNA in slight RNA excess; to this hybridization mixture is added unlabeled sample RNA, and the extent to which the normal incorporation of RNA into duplex is inhibited is measured. By use of a calibration curve, a quantitative estimate of the concentration of the mRNA sequences present in an RNA sample can be made, an estimate that theoretically can be fairly sensitive, since very high specific activity RNA can be obtained by in vitro labeling techniques [36, 37] . Poly-(A) sequences are not required, and the whole mRNA sequence is involved. With histone mRNA, the method is particularly attractive because amplification of the histone gene [38, 39] makes the DNA sequences required for the assay fairly easy to obtain. Obtaining enough DNA corresponding to mRNAs transcribed from unique sequences is a problem that deserves experimental attention. Also, this method does not easily allow an estimation of the rate at which radioactive label flows through the measured mRNA species.

## III. APPLICATIONS

### A. Kinetics of mRNA Metabolism

The ability to assay for specific mRNA sequences suggests experiments that measure the basic kinetic parameters of mRNA synthesis. The general protocol for such a series of experiments involves the isolation of the total RNA in a cell or tissue at intervals during a period of specific mRNA accumulation, such as during cell differentiation. Analysis of the kinetics of hybridization of highly labeled cDNA to known quantities of those RNA samples yields a measurement of both the relative and absolute rates of mRNA accumulation. Preliminary measurements of this kind have been performed with globin cDNA in mouse cell lines infected with Friend leukemia virus [40], which synthesize globin after treatment with dimethylsulfoxide [41] . Globin mRNA accumulates after a lag of about a day after $Me_2SO$ induction and reaches a plateau after approximately three days of incubation. Further experiments are required before any particular model for mRNA accumulation [42] can be either entertained or eliminated on the basis of this kinetic evidence. In order to completely and definitively characterize the kinetics of mRNA accumulation, synthesis and decay must be differentiated experimentally, by study of the entrance and exit of radioactive label into mRNA, selectively isolated from a total mRNA mixture.

But this experiment would require an excess of DNA so that the mRNA could be quantitatively brought into hybrid form and its specific radioactivity measured as a function of time of labeling or chase. Difficulties in obtaining large quantities of cDNA have thus far precluded investigations of this sort.

### B. Distribution and Movement of mRNA Sequences

The ability to probe for mRNA sequences is not restricted to samples of total RNA. By studying RNAs isolated from subcellular fractions rather than from whole cells, information can be obtained on the flow of specific mRNA sequences into and through such cell compartments as the cell sap or membrane-bound polysomes. An interesting extension of this study of mRNA localization is the possibility of using in situ hybridization techniques. By analogy with the studies of Pardue and Gall [43], it may be feasible to hybridize labeled cDNA to mRNA sequences in tissue sections and subsequently to localize the hybridizing species by autoradiography. Although as yet unexploited, these methods will bring under scrutiny the spatial aspects of both sub- and supercellular patterns of gene regulation, complementing the study of the temporal aspects embodied in the kinetic approaches outlined above.

## C. mRNA Transcription

A wide variety of evidence from bacterial, as well as from animal cells has focused attention on the possibility that regulation may occur at the level of transcription of the genome into informational RNA. In eukaryote cells, the DNA template appears to exist only as a very complex structure in combination with protein and RNA, whose in vitro incarnation is termed chromatin[44]. The existence of this complex structure emphasizes a primary and fundamental question regarding transcriptional regulation, namely, whether regulatory factors act by means of an interaction with the DNA itself or with the RNA polymerase that transcribes it. Clearly, the elucidation of the contribution from each of these alternatives to the regulation of mRNA synthesis will require an ability to analyze the transcription process in vitro. Considering the complexity of the structures involved, the first steps toward the unraveling of these problems that have already taken place are nothing short of remarkable. An important first experiment was the use of E. coli RNA polymerase, which presumably is insensitive to most eukaryotic regulatory factors (if such exist) to synthesize RNA on a chromatin template isolated from immature erythroblasts [45, 46]. Examination of the transcript from this reaction with the cDNA hybridization assay indicated that globin mRNA sequences were among those transcribed and were present in concentrations considerably higher than their representation in the DNA itself. These results strongly imply that the structure of the DNA-protein complex is itself capable of directing the synthesis of certain gene products over others, irrespective of the (regulatory) status of the polymerases. The probability that the polymerase also plays a role in regulation is indicated by experiments suggesting that the yield of globin sequences is augmented by the use of a mammalian RNA polymerase instead of the bacterial enzyme [47]. Analysis of this area, although clearly still in its infancy, already promises to yield exciting results.

### D. Specific Gene Amplification

Analogy with the case of rRNA in oocytes [48] raises the possibility that regulation of gene expression may occur by selective amplification of expressed genes in cells differentiated to express those genes. This possibility is related to another; namely, that structural genes are more or less generally amplified in somatic cells as a key to gene specialization [49]. A test of these proposals is possible using cDNA as a tracer for the specific expressed gene in a DNA renaturation experiment, in which reiteration frequencies can be estimated from the kinetics of duplex formation. In the ideal case, the cDNA would be so highly labeled that it could be included in a DNA renaturation at levels considerably lower than those contributed by the genome itself. In practice, the cDNA can make a significant contribution to the concentration of specific sequences, but the correction for this effect is straightforward and has been described by Bishop and Freeman [50]. These experiments can use labeled mRNA molecules as the specific probe [51], but cDNA has the advantage in that there is no requirement for any correction for differences in the rate at which RNA, compared to DNA, enters the duplex form. Thus, inclusion of cDNA in renaturation experiments allows an easily interpretable method of estimating the reiteration frequency of specific genes in DNA from cells highly differentiated in the production of the specific mRNA in question. This experiment has been performed [25,50,52-54]. With globin, the number of genes coding for the specific protein was originally estimated to be surprisingly high, about five to sixteen copies [53], but the number has fallen steadily since, and is almost certainly less than three [50]. All of the evidence, moreover, has consistently implied that whatever the number of copies, there is no difference between cells which do and those which do not synthesize the specific mRNA [52-54].

## ACKNOWLEDGMENTS

This work was supported by a USPHS Grant (No. 15885) and by an N.I.H. Grant (No. HD-01229).

## REFERENCES

1. H. Latham and J. E. Darnell, J. Mol. Biol., 14, 13 (1965).
2. S. Penman, C. Vesco, and M. Penman, J. Mol. Biol., 34, 49 (1968).
3. H. M. Temin and S. Mizutani, Nature, 226, 1211 (1970).
4. D. Baltimore, Nature, 226, 1209 (1970).
5. I. M. Verma, F. Temple, H. Fan, and D. Baltimore, Nature (New Biology), 235, 163 (1972).
6. D. L. Kacian, S. Spiegelman, A. Bank, M. Terada, S. Metafora, L. Dow, and P. A. Marks, Nature (New Biology), 235, 167 (1972).
7. J. Ross, H. Aviv, E. Scolnik, and P. Leder, Proc. Natl. Acad. Sci., U.S.A., 69, 264(1972).
8. Patrick Williamson, Ph.D. thesis, Harvard University, 1974.
9. R. Williamson, M. Morrison, G. Lanyon, R. Eason, and J. Paul, Biochemistry, 10, 3014 (1971).
10. W. D. Sutton, Biochim. Biophys. Acta, 240, 522 (1971).
11. V. M. Vogt, Eur. J. Biochem., 33, 192(1973).
12. H. D. Preisler, D. Housman, W. Scher, and C. Friend, Proc. Nat. Acad. Sci., U.S.A., 70, 2956(1973).
13. B. D. Young, and J. Paul, Biochem. J., 135, 573(1973).
14. R. J. Britten and D. E. Kohne, Science, 161, 529,(1968).
15. J. O. Bishop, Gene Transcription in Reproductive Tissue, E. Dicfalusy, ed. Karolinska Institutet, Stockholm, Vol. 5, 1972, p. 247.

16. J. Paul and R. S. Gilmour, J. Mol. Biol., 34, 305 (1966).

17. J. Paul and R. S. Gilmour, J. Mol. Biol., 34, 305 (1968).

18. J. O. Bishop, Biochem. J., 113, 805(1969).

19. J. O. Bishop, Biochem. J., 116, 223(1970).

20. I. M. Verma and D. Baltimore, Methods in Enzymology, L. Grossman and K. Moldave, eds. Vol. 29, Academic Press, New York, 1973, p. 125.

21. H. Diggelman, C. H. Faust, Jr., and B. Mack, Proc. Natl. Acad. Sci., U.S.A., 70, 693(1973).

22. J. O. Bishop and K. B. Freeman, Cold Spring Harbor Symp. Quant. Biol., 38, 707(1973).

23. P. Leder, J. Ross, J. Gielen, S. Packman, Y. Ikawa, H. Aviv, and D. Swan, Cold Spring Harbor Symp. Quant. Biol., 38, 753(1973).

24. J. B. Clegg, D. J. Weatherall, and P. F. Milmer, Nature, 234, 337(1971).

25. D. Sullivan, R. Palacios, J. Stavnezer, J. M. Taylor, A. J. Faras, M. L. Kiely, N. M. Summers, J. M. Bishop, and R. T. Schimke, J. Biol. Chem., 248, 7530(1973).

26. H. Aviv, S. Packman, D. Swan, J. Ross, and P. Loder, Nature(New Biology), 241, 174(1973).

27. P. M. Lizardi and D. D. Brown, Cold Spring Harbor Symp. Quant. Biol., 38, 701(1973).

28. M. Melli and R. E. Pemberton, Nature (New Biol., 236, 172(1972).

29. M. S. Robert, R. G. Smith, R. C. Gallo, P. S. Sarin, and J. W. Abrell, Science, 176, 798(1972).

30. J. M. Taylor, A. J. Faras, H. E. Varmus, H. M. Goodman, W. E. Levinson, and J. M. Bishop, Biochemistry 12, 460(1973).

31. J. D. Karkas, J. G. Stavrianopoulos, and E. Chargaff, Proc. Natl. Acad. Sci., U. S. A., 69, 398(1972).

32. L. A. Loeb, K. D. Tartof, and E. C. Travaglini, Nature(New Biol.),242, 66(1973).

33. M. J. Modak, S. L. Marcus, and L. F. Cavalieri, Biochem. Biophys. Res. Comm., 55, 1(1973).

34. S. C. Gulati, D. L. Kacian, and S. Spiegelman, Proc. Natl. Acad. Sci., U.S.A., 71, 1035(1974).

35. A. Skoultchi and P. R. Gross, Proc. Natl. Acad. Sci., U.S.A., 70,2840(1973).

36. A. Tereba and B. J. McCarthy, Biochemistry, 12, 4675(1973).

37. N. H. Scherburg and S. Refetoff, Nature(New Biol.), 242, 142(1973).

38. L. H. Kedes and M. L. Birnstiel, Nature(New Biol.), 230, 165(1971).

39. B. J. McCarthy, J. T. Nishiura, D. Docnecke, D. S. Nasser, and C. B. Johnson, Cold Spring Harbor Symp. Quant. Biol., 38, 763(1973).

40. J. Ross, Y. Ikawa, and P. Leder, Proc. Natl. Acad. Sci., U.S.A., 69, 3620(1972).

41. C. Friend, W. Scher, J. G. Holland, and T. Sato,

Proc. Natl. Acad. Sci., U.S.A., 69, 3620(1972).

42. F. C. Kafatos, Gene Transcription in Reproductive Tissues, E. Diczfalusy, ed. Vol. 5, Karolinska Institutet,Stockholm , 1972, p. 319.

43. M. L. Pardue and J. G. Gall, Science, 168, 1356(1970).

44. D. E. Comings, 1972, Advances in Human Genetics, H. Harris and K. Hirshhorn, eds. Vol. 3, Plenum Press, New York, p. 127.

45. R. Axel, H. Cedar, and G. Felsenfeld, Proc. Natl. Acad. Sci., U.S.A., 70, 2029(1973).

46. R. S. Gilmour, and J. Paul, Proc. Natl. Acad. Sci., U.S.A., 70, 2440(1973).

47. A. W. Steggles, G. N. Wilson, J. A. Kantor, D. J. Picciano, A. K. Falvey, and W. F. Anderson, Proc. Natl. Acad. Sci., U.S.A., 71, 1219(1974).

48. I. B. Dawid, D. O. Brown, and R. H. Reeder, J. Mol. Biol., 51, 341(1970).

49. H. G. Callan, J. Cell Sci., 2, 1(1967).

50. J. O. Bishop and K. B. Freeman, Cold Spring Harbor Symp. Quant. Biol., 38, 707(1973).

51. M. Melli, C. Whitfield, K. V. Rao, M. Richardson, and J. O. Bishop, Nature(New Biol.), 231, 8(1971).

52. P. R. Harrison, A. Hell, G. D. Birnie, and J. Paul, Nature, 239, 219(1972).

53. S. Packman, H. Aviv, J. Ross, and P. Leder, Biochem. Biophys. Res. Commun., 49, 831(1972).

54. J. O. Bishop and M. Rosbash, Nature(New Biol.), 241, 204(1973).

Chapter 3

# ANALYSIS OF mRNA TRANSCRIPTION IN *DICTYOSTELIUM DISCOIDEUM* OR SLIME MOLD MESSENGER RNA: HOW TO FIND IT AND WHAT TO DO WITH IT ONCE YOU'VE GOT IT

Allan Jacobson
Department of Microbiology
University of Massachusetts Medical School
Worcester, Massachusetts

## INTRODUCTION

In recent years, considerable effort has been expended on the characterization of messenger RNA (mRNA) metabolism in the cellular slime mold, *Dictyostelium discoideum* [1-8]. It has been shown that Dictyolstelium mRNA contains polyadenylic acid [poly (A)] sequences [1,2,6], a property shared with virtually all other eukaryotic mRNA molecules [9,10]. This has permitted the purification and analysis of such molecules by hybridization with immobilized poly(U) or oligo(dT) [1,2] . Further, the majority of pulse-labeled *Dictyostelium* nuclear RNA that is not precursor to ribosomal RNA (rRNA) has been shown to also contain at least one poly(A) sequence, and this RNA appears to be only 20% larger than its cytoplasmic counterpart [2]. Seventy percent of the nuclear poly(A)-containing RNA is transferred to polysomes and, hence, is a material precursor to mRNA [2,4]. This mRNA precursor, as well as the rRNA precursor, can be synthesized in isolated nuclei incubated in vitro with the four ribonucleoside triphosphates [5]. Studies on RNA synthesis in isolated nuclei have shown that *Dictyostelium* mRNA contains two size classes of poly(A): one approximately 25 nucleotides long and one approximately 100 nucleotides long [5,6]. The former is coded for by $(dT)_{25}$ sequences in DNA, whereas the latter is added post-transcriptionally [6]. All of the above results, as well as data on DNA renaturation kinetics [11], have been integrated to provide a model for the structure of an "average" portion of the *Dictyostelium* genome [4,7,8].

The methods used in all of these experiments are discussed below.

## II. GROWTH, DEVELOPMENT AND STORAGE OF CELLS

### A. Growth of Amoeba

The axenic strain, AX-3, is used in all experiments. The medium used for growth of vegatative AX-3 amoeba is basically the HL5 medium described by Watts and Ashworth [12] and Cocucci and Sussman [13], in which the phosphate buffer has been replaced by MES buffer. Although this modification was made specifically to facilitate labeling with radioactive phosphate, it is now routinely used as our growth medium because cells grown in it show the same doubling time and developmental capacities as cells grown in standard HL5 medium.

MES-HL5 (Per Liter)

5 g Difco yeast extract
10 g Difco proteose peptone
10 g glucose
1.3 g MES [2-(N-morpholino) ethane sulfonic acid, monohydrate]

Adjust to pH 6.6 with 5M KOH, distribute to flasks according to the schedule below and autoclave fifteen minutes.

We have not encountered problems with "caramelizing" of glucose during autoclaving. Ordinary distilled water is fine. We have used other sources of yeast extract and proteose peptone, e.g., BBL, but have found the Difco products to be more reliable in the long run. However, since some batches of Difco media do prove unsatisfactory for slime mold growth, keep track of lot numbers. The most commonly occurring strange thing that happens to our media is the accumulation of dark red algae-like flakes during growth. These appear to be nonliving matter that have no effect on *Dictyostelium*.

To ensure the proper aeration, the following volumes of medium are used in the flasks of designated size:

| Flask Capacity | Volume of MES-HL5 |
|---|---|
| 125 ml | 35 ml |
| 250 ml | 80 ml |
| 500 ml | 140 ml |
| 1 liter | 340 ml |
| 2 liter | 800 ml |
| 6 liter | 0140 ml |

We routinely prepare large batches of medium and store it at room temperature for up to several months. There is no loss in "potency" of the medium during this time.

Cultures are grown on a rotary platform shaker cycling at 130 to 150 rpm at 19 to 23$^{o}$ C. For optimal growth, we prefer to maintain the cells at 22 to 23$^{o}$ C. In a building that is well air-conditioned, room temperature is often perfect, but risky. Cells in smaller flasks (less than 2 liter) grow better at 150 rpm. Cells in larger flasks grow better at 130 rpm. Cultures are not being shaken fast enough when white rings of cells accumulate inside a flask. It will be found that cells will grow with reasonable efficiency at other shaker speeds, e.g., faster than 150 rpm. However, such cells generally do not develop properly.

Cell growth is routinely monitored by counting in a hemocytometer. Cell densities of up to 2 X $10^7$ cells/ml can be attained but should be avoided, although cells are happily growing logarithmically up to densities of 5 x $10^6$ cells/ml.

## B. Development

Most of the technology of this section has been described in detail by Sussman [14]. For completeness, we describe our variations of his basic method.

### Preparation of Cells

The term development is loosely used here to describe the morphogenetic events that occur when cells are removed from growth medium and placed on a solid substratum. The operations involved are the following:

Cells growing in MES-HL5 are harvested by centrifugation at 4$^o$ C. For proper development, cells should be harvested at densities of less than 5 x $10^6$/ml. When a Sorvall centrifuge is used, cells are pelleted at 1500 rpm for five minutes in the SS34 rotor or at 2000 rpm for five minutes in the GSA or the GS-3 rotor. Cell pellets are washed with sterile ice-cold MES-LPS (see below) and recentrifuged. This operation is repeated once. Although MES-LPS is routinely used to wash cells, 0.2% NaCl works just as well. The final cell pellet is resuspended in MES-LPS at a final concentration of 1.6 x $10^8$ cells/ml. This concentration can be determined by counting an appropriate dilution or by approximating an 80% recovery of the original population of cells. These cells are ready to be placed on filters for development and should be kept on ice until that time. We find that storage on ice for more than fifteen minutes substantially impairs development. Therefore, it is imperative that the filters and plates for developing cells be prepared in advance, so that they may be used as soon as the cells are harvested, washed, and resuspended.

MES-LPS (Per Liter)

1.5 g MES
1.5 g KCl
0.6 g $MgSO_4$

Autoclave twenty minutes, cool, then add 10 ml per liter of filter-sterilized streptomycin sulfate. Store at $4^{o}$ C.

Preparation of Filters

The solid substratum on which cells develop is a Whatman #50 filter. This rests on a filter pad inside a plastic Petri dish. We routinely use two sizes of Petri dish, 60 x 15 mm ("small") and 150 x 15 mm ("large"). The lower pad used in small Petri dishes is a sterile Millipore AP 10047S0 filter. A 4.25 cm Whatman #50 filter circle rests on top of this. In large Petri dishes, the lower pad is a 12.5 cm Whatman #3 filter circle. Resting on this is a 12.5 cm W & R Balston Whatman #50 filter circle.

To prepare plates and filters for development, place the appropriate combination of pad and filter in a Petri dish of the desired size. Add 4 ml MES-LPS to a small dish (and filter) and 12 to 13 ml to the larger. Let the solution soak in (at room temperature) for at least five minutes. Fanatic sterility is not required. Turn the small dishes on their sides and pipette off excess liquid. Excess is anything that forms a droplet on the side of the pad. Do the same for the large dishes. Even more excess liquid can be removed from a large dish by "squeezing" it with a glass spreader. The lower halves of the dishes are now ready.

To prepare the upper halves of the dishes, tape the "pad" moiety of the previous combination, i.e., Whatman #3 or Millipore AP 10047S0 to the upper part of the appropriate dish. Use <u>one</u> piece of tape. Saturate this pad with UPS (see below), let it soak in, and remove the excess as described above.

UPS (Per Liter)

Prepare 1 M $Na_2HPO_4$
1 M $KH_2PO_4$

Mix at 1:5 ($Na_2$:K) to pH 6
Autoclave 15-20 minutes
Store at $4^o$ C.

Both halves of the Petri dish are now ready. They may remain at room temperature for several hours before use.

Plating Cells on Filters

With the filters and dishes and, subsequently, the washed cells prepared everything is ready to be combined. On a small filter place 0.33 ml of washed and resuspended cells; on a large filter place 5-6 ml of the same. Keep swirling the flask of cells to keep them from settling out. Place the cells on the filters starting at the center of the filter and proceeding in a spiral fashion to the periphery. With the covers off, let cells "soak in" at room temperature. Allow five minutes for small filters and fifteen minutes for large filters, then tilt the plates and remove the excess moisture. Then, close the lids, wrap dishes in Saran Wrap, and incubate at $22^o$ C in visible light. Dishes are not wrapped individually, but in stacks. Small dishes can often be stacked inside a baking dish and then wrapped in Saran Wrap.

The morphologic changes characteristic of the twenty-four to twenty-six hours of slime mold development have been described in detail elsewhere [15, 16, 17] and the reader is directed to use these references as guides in the identification of different stages. It is not uncommon to discover that cells are not developing according to the prescribed schedule or that aberrant structures accumulate (the most common of the latter being "fingers"). Such abnormal development may be attributed to any number of problems. Often, these include excessive or insufficient moisture, low cell density on filters, or improperly buffered MES-LPS. Occasionally, it is also necessary to change the lower pads at approximately twelve hours of development. This procedure apparently counteracts the effect of accumulated extracellular material that inhibits development. If none of the above ills can be shown to have occurred, it is routine to go back to stocks and begin with a new clone of AX3. It is common to find that AX3 grown in continuous culture will lose the ability to develop properly after four to six weeks. To avoid unnecessary time loss, several different clones of AX3 should be maintained at all times.

## C. Storage of Cells

### Short-Term Storage

For short-term storage, amoeba (or spores) can be plated on SM agar in association with *Aerobacter aerogenes* (*Klebsiella*) or *E. coli*. Thus, 0.2 ml of an overnight bacterial culture is mixed with an appropriate dilution of amoeba or spores and spread on an SM plate. Alternatively, the bacteria are spread on the plate and allowed to soak in, and a streak of Dictyostelium amoeba is then spread across the surface. In either case, plaques will arise in the bacterial lawn at sites where individual amoeba (or spores) were originally deposited. Fruiting bodies, apically bearing as many as $5 \times 10^4$ spores, will form within these plaques. These spores can be picked up with a sterile needle and reinoculated into MES-HL5. Such agar plates can be used to store spores for several months at $4^o$ C.

SM Agar 14 /liter

10 g glucose
10 g Difco Bacto peptone
1.0 g $MgSO_4 \cdot 7H_2O$
2.2 g $KH_2PO_4$
1.0 g $K_2HPO_4$
1.0 g Difco yeast extract
20 g Difco Bacto agar

Autoclave fifteen minutes.

SM broth (for cultivation of bacteria) is the same as SM agar minus Bacto agar.

Long-Term Storage

For long-term storage of Dictyostelium, spores are lyophilized, or are stocked on silica gels, or whole cells are frozen. To harvest spores, a mature plate of cells developed on SM agar is inverted and tapped. The spores fall into the lid. These can be collected into a few milliliters of cold sterile 5% Starlac (instant milk) and stored on silica gels. In the latter procedure [18], about 1 cm of color-free silica gel is sterilized (dry heat) in a one to two dram glass vial (cover off). Chill the vial on ice, add about 0.5 ml of Starlac and spores, recap, shake vigorously, and put back on ice. Store, dessicated, at room temperature. To recover cells from spores, add lyophilized spores, or silica gel to some SM broth, prepare dilutions, and plate with bacteria on SM agar. Spores can then be picked up from fruiting bodies and reinoculated into MES-HL5.

An alternative to storage of spores is the freezing of whole cells [19]. Begin by sterilizing $(Me)_2SO$ by heating it at 80° C for fifteen minutes. Then, harvest a log-phase culture and wash twice with cold MES-HL5. Resuspend at 0.5-5 x $10^7$/ml in cold MES-HL5. Add $Me_2SO$ to a final concentration of 10%. Transfer 0.5 ml to prechilled sterile 0.5 dram vials and store at -70° C. When thawing, swirl constantly, then dilute and plate. Survival is 0.1-1%.

## II. LABELING WHOLE CELLS WITH RADIOISOTOPES

### A. Amoeba

Amoeba are labeled merely by adding a given isotope to the MES-HL5 medium. Labeling is terminated by centrifuging the cells from the medium, rapid chilling, or both. Precursors to RNA and DNA, e.g., $^{32}PO_4$, thymidine, uridine, uracil, adenine, and adenosine, are readily incorporated in MES-HL5. Amino acids, however, are not readily incorporated in this medium because of its high content of peptide and "cold" amino acids. Incorporation of $^{35}SO_4$ into protein is also not efficient in MES-HL5.

We have found that the least expensive, and most useful, label for most experiments on RNA or DNA has been $^{32}PO_4$. We use it in its HCl-free form. To label rRNA or DNA, the $^{32}PO_4$ is included in the culture medium for several generations. Messenger RNA requires only a fifteen to sixty minute label and mRNA precursor requires but a two-minute labeling [1,2] . For all of these labelings, we routinely use ten to one hundred mCi of $^{32}PO_4$ in approximately 140 ml of culture. All labelings are done with log-phase cells, i.e., less than 5 x $10^6$/ml, although for very short labelings we have had considerable success with cells that were tenfold concentrated. We have not found that there is a corresponding increase in specific activities when the size of cultures has been diminished. In general, we have been dissatisfied with the labeling efficiency in very small cultures.

Although it should be common knowledge, it is imperative to point out that the use of such large quantities of $^{32}P$ constitutes a radiation hazard and that adequate precautions must be taken to protect lab personnel and to appropriately dispose of waste. With regard to protection, it must be noted that the use of thin lead film for shielding is more dangerous than direct exposure to the $^{32}P$. Bombardment of the lead by the β-rays of $^{32}P$ gives rise to far more lethal heavy metal ionizations. A safe shielding for $^{32}P$ is thick Plexiglas, or some related plastic product.

Given the radiation hazard of $^{32}P$, it can be asked how significant are its advantages? First of all, as noted above, it is the least expensive RNA or DNA precursor. For example, in a recent experiment, the ratio of dollars to counts per minute in mRNA was $10^{-7}$. Further, with $^{32}P$, all stages of an experiment can be monitored with a Geiger counter or a scintillation counter ($^{3}H$ channel) to count Cerenkov radiation. One also obtains uniform labeling of all nucleotides and high specific activity, thus permitting sequence analysis. The high energy of the isotope also allows for autoradiography, even through thick slabs of polyacrylamide (see below).

### B. Developing Cells

Developing cells are labeled by either of two ways--the label is either added to the lower pad or to the upper filter alone. Isotope is added to the lower pad when labeling times exceed one to two hours. Usually the isotope is added to a new pad, and the Whatman #50 filter containing cells is transferred to this pad when labeling is desired. Labeling is terminated by washing the cells off the filter or by a second transfer to a label-free pad. For short labeling periods, the Whatman #50 filters containing developing cells are transferred directly to droplets of isotope resting in the bottom of Petri dishes. Labeling is terminated as above. Uracil, uridine, adenine, adenosine, and amino acids label efficiently on small or large filters. Thymidine is not incorporated in development because there is no DNA synthesis. The $^{32}PO_4$ (10-50 mCi/plate ) is incorporated into RNA more efficiently on large filters. "Developmental" rRNA is usually labeled from two or three hours to fifteen to nineteen hours of development, whereas mRNA is labeled efficiently in developing cells in thirty to 180 minutes, and its precursor in five to fifteen minutes. There is an apparent delay in the processing of mRNA and 18s rRNA during development [20]. This gives rise to the labeling discrepancies in amoeba and in developing cells.

## III. PREPARATION OF SUBCELLULAR FRACTIONS

The subcellular fractions pertinent to this discussion are nuclei, cytoplasm, and polysomes.

### A. Isolation of Nuclei

Nuclei are isolated by differential centrifugation of detergent-lysed cells by a modification of the procedure of Cocucci and Sussman [13]. Amoeba are harvested by centrifugation at 500 g for five minutes and then washed twice with 0.2% NaCl. Developing cells are washed off the filter with 0.2% NaCl, centrifuged, and then washed one additional time with NaCl. Cell pellets are resuspended in ten to twenty volumes of ice-cold lysis buffer (see below), and vortexed at $4^{o}$ C for forty-five to sixty seconds. Lysis of late-developing cells (later than about sixteen hours) may require, in addition, ten to fifteen strokes in a Dounce homogenizer. Debris and unbroken cells are removed by centrifugation at 400 g for five minutes, and the supernatant is then centrifuged at 2000 g for five minutes. The resulting nuclear pellet is again resuspended in lysis buffer, vortexed, and centrifuged at 2000 g. For ultra-pure nuclei (used for in vitro transcription), the latter steps are repeated three times. For isolation of labeled nuclear RNA or DNA, nuclei at this stage of purity are usually adequately free of cytoplasmic contamination. With either procedure, a 400 g centrifugation preceding the last extraction efficiently removes the remaining whole cell contaminants.

The purity of nuclear preparations (and the efficiency of lysis) is monitored by phase-contrast microscopy; contamination with unbroken cells is usually less than 0.01%.

Extraction of DNA or RNA from nuclei is discussed in another section. For in vitro transcription, the final nuclear pellet is resuspended in two to five volumes of storage buffer (see below), distributed into 100-ml aliquots, quick-frozen in a dry-ice-ethanol bath, and stored at $-70^{\circ}$ C. Activity is stable at this temperature for twelve to fifteen months at least. Moreover, DNA isolated from stored nuclei appears to have the same single-strand size as DNA isolated from fresh nuclei.

Lysis Buffer

| | |
|---|---|
| 0.05M | HEPES (pH 7.5) |
| 5 mM | $MgAc_2$ |
| 10% | Sucrose |
| 2% | Cemusol NPT12 |

This buffer contains an "overkill" amount of detergent. But, there are rarely any problems encountered with incomplete cell lysis.

Storage Buffer

| | |
|---|---|
| 0.04 M | Tris-HCl (pH 7.9) |
| 0.01 M | $MgCl_2$ |
| 0.1 mM | $Na_2$ EDTA |
| 1 mM | Dithiothreitol |
| 50% | Glycerol |

B. Isolation of Cytoplasm

Cells are harvested and washed as in the protocol for nuclear isolation (above). Cell pellets are again resuspended in ten to twenty volumes of ice cold lysis buffer (see above) and vortexed at $4^{o}$ C for forty-five to sixty seconds.

Debris, unbroken cells, mitochondria, nuclei, and other vesicular structures are removed by centrifugation at 20,000 g for fifteen minutes. The resulting supernatant is defined as cytoplasm.

Cells can be lysed in the absence of detergent by first resuspending them in 0.01M HEPES (pH 7.5), 0.01M KCl, and 3 mM dithiothreitol and incubating ten minutes at $4^{o}$ C. Subsequently, fifteen to twenty strokes of a tight Dounce homogenizer will suffice to lyse the cells. The lysate may then be centrifuged as above to generate detergent-free cytoplasm.

## C. Isolation of Polysomes

Polysomes are isolated from cytoplasm of detergent-lysed cells by centrifugation to the bottom of a tube, to a sucrose cushion, or throughout a sucrose gradient.

Polysomes are pelleted by overnight centrifugation through a cushion of 1M sucrose in 0.05M HEPES (pH 7.5), and 5 mM $MgAc_2$, at $4^{o}$ C. We usually use a Spinco SW27 rotor and spin at 27,000 rpm. Polyallomer tubes are half-filled with sucrose solution and overlaid with lysate. The polysomal pellet is the jelly-like thing at the bottom of the tube.

Where bulk and speed are important, polysomes are sedimented onto a cushion of 1.7M sucrose, also in 0.05M HEPES (pH 7.5), 5 mM $MgAc_2$. In the SW27 rotor, 5 ml of 1.7M sucrose are overlaid with ten to twenty ml of 1M sucrose, and the tube is topped off with lysate. After centrifuging at 27,000 rpm at $4^{o}$ C for six to seven hours, once can recover the white "band" of polysomes at the interphase of the two sucrose solutions. This is done by aspirating away most of the material above the band and then inserting a Pasteur pipette into it.

To resolve size classes of polysomes, the cytoplasmic extract is layered over a 15 to 50 percent sucrose gradient in 0.01M HEPES (pH 7.5), 0.01M $MgCl_2$, and 0.01M KCl. Centrifugation in the SW41 rotor, at $4^{o}$ C, is for 120 to 135 minutes at 41,000 rpm. Centrifugation in the SW27 rotor, at $4^{o}$ C, is for four to five hours at 27,000 rpm.

We have recently found that better profiles of polysomes are obtained if cycloheximide (500 μg/ml) is added to cells immediately before harvesting and also included in the cell wash and the lysis buffers. Presumably, the drug is inhibiting ribosome run-off.

## IV. ISOLATION OF RNA

### A. Phenol Extraction

The RNA is isolated from cells or subcellular fractions by extraction with a mixture of phenol, chloroform, and isoamyl alcohol [1,2,5]. Washed cell pellets are resuspended in ice cold 0.05M Tris (pH 7.5), at a final concentration of about five to ten x $10^7$ cells/ml. Cells are lysed by adding SDS to a final concentration of 0.5%, followed by vortexing. To inhibit nucleases, diethylpyrocarbonate is immediately added to a final concentration of 1%, while mixing is continued. One and one-half volumes of a cold mixture of phenol-chloroform-isoamyl alcohol (66:33:1) is subsequently added and vigorous shaking is continued for one to five minutes. Aqueous and organic phases are separated by centrifugation at 12,000 g for ten minutes. The aqueous phase is reextracted with the phenol-chloroform-isoamyl alcohol mix at least three times, or until there is no longer any detectable material at the interphase found after centrifugation.

The RNA is isolated from nuclei or pelleted polysomes by resuspending them in reduced volumes of 0.05M Tris (pH 7.5), and proceeding as above. Fewer repetitive extractions are required here. The RNA is isolated from polysomes in sucrose by adding SDS and diethylpyrocarbonate to the sucrose solution and extracting with phenol. Beware of an inversion of phsses after centrifugation if the aqueous phase contains a lot of sucrose.

We have found that the use of diethylpyrocarbonate significantly impairs degradation of slime mold RNA during isolation. In control experiments with bacteriophage f2 RNA, we have shown that the RNA isolation conditions described here do not impair the ability of that RNA to direct in vitro protein synthesis [21]. Moreover, we have noted that the use of diethylpyrocarbonate at elevated temperatures, e.g., 20 - 37$^{o}$ C, did alter the f2 RNA, in that both initiation and elongation of polypeptide chains was reduced relative to an untreated control or to an RNA treated at 4$^{o}$ C.

We routinely redistill our phenol, saturate it with water, and add 1% 8-quinolinol as a preservative. Phenol solutions in immediate use are stored in the dark at 4$^{o}$ C. Extra phenol is stored in the dark at -20$^{o}$ C.

B. Ethanol Precipitation

After phenol extraction, RNA is precipitated by the addition of 0.2 volume of 2M NaAc and 2.5 volumes of chilled 95% EtOH, followed by centrifugation for one to sixteen hours in any of several ultracentrifuge rotors. Centrifugation is at 15-17$^{o}$ C at top rotor speed in polyallomer tubes. Under these conditions, recovery of RNA is quantitative when more than 100 μg of RNA is being precipitated. To quantitatively recover very small quantitites of RNA we use a procedure established by Roberts [22]. In this procedure 0.1 volume of 1M $Na_2HPO_4$ (equimolar solution of mono- and dibasic salts) is added instead of NaAc. After the addition of 2.5 volumes of 95% EtOH, all the $NaHPO_4$ precipitates, with all the RNA. The precipitate is collected by centrifugation at 12,000 g for ten minutes, dried in a lyophilizer for ten to fifteen minutes, and then resuspended in water. The RNA is separated from the salt by G-25 Sephadex chromatography, usually in a Pasteur pipette. The RNA comes through in the excluded volume and is subsequently freeze-dried in siliconized tubes. Small quantities of RNA, e.g., less than 50 μg, can thus be quantitatively recovered and ultimately resuspended in volumes as small as 1 μl.

## V. ISOLATION OF NUCLEAR DNA

Slime mold nuclear DNA is isolated by a modification of the procedure of Firtel and Bonner [11]. Nuclei are purified as described in Section III. The final nuclear pellet is resuspended in 0.2M EDTA (pH 8.0), at approximately $10^9$ nuclei/ml. This suspension is adjusted to a final concentration of 4% N-lauroyl-sarcosine (using a 20% stock solution), swirled, and heated at 55 to 60° C for four to five minutes (or until the solution clears). Subsequently, 0.9 g of CsCl/ml of suspension is quickly added; swirling and heating at 55 to 60° C is continued until all of the CsCl is dissolved and then for five minutes more. After dissolving the CsCl, cool the solution to room temperature, and add 1 ml of a 10 μg/ml stock of ethidium bromide/20 ml of solution. Stir constantly during the addition of ethidium bromide. Transfer everything to a screw-cap polypropylene ultracentrifuge tube. Fill to the top with a balancing solution of CsCl ($\rho$ = 1.57 g/ml = 0.95 g CsCl/ml of $H_2O$ added). Centrifuge in a Spinco type 30 rotor for sixty hours at 28,000 rpm at 40° C, or in a type 50 Ti rotor at 45,000 rpm for forty-eight hours at 4° C. After centrifugation, the DNA will be in a red band, roughly in the middle of the tube. On top will be a red protein skin. Remove this with forceps, aspirate off the CsCl to within two to three cm of the DNA, insert a wide-bore pipette into the DNA band, and suck it out. Ordinarily, we rerun the DNA in a second gradient, as above. The final DNA "bands" are pooled and stored, as is, in the dark at 4° C. This is advised because ethidium bromide apparently acts as a free radical sink [23], preventing damage to the DNA. When desired, ethidium bromide and CsCl are removed as follows: dilute the stored DNA with two volumes of distilled water and pass through a small (about 1 ml) column of Dowex 50-X8 (that has been previously adjusted to pH 8). Collect the flow-through and one volume of wash water. Then dialyze against 0.2M NaAc, 1 mM EDTA for two hours, followed by dialysis against 0.2M NaAc alone. Store at $4^{\circ}$ C.

## VI. FRACTIONATION OF RNA

In this section, methods are described for (a) separating mRNA from rRNA by affinity chromatography on poly(U)-Sepharose, poly(U) filters, and (dT)-cellulose, and (b) fractionating any RNA on acrylamide gels or sucrose gradients.

### A. Preparation of Poly(U) Filters and Poly(U)-Sepharose

Poly(U) filters are prepared by the method of Sheldon et al. [24]. Place GF/C filters (2.4 cm) on a sheet of parafilm and spot 100 μl of a 1 mg/ml solution of poly(U) in the center two-thirds of the filter. Dry overnight at $37^{\circ}$ C and then irradiate for two to four minutes/side at 15 cm from a 15-W or 22 cm from a 30-W germicidal UV lamp. Store at $-20^{\circ}$ C.

Poly(U)-Sepharose is prepared from poly(U) and cyanogen bromide-activated Sepharose 4B by a modification of the methods suggested by the manufacturer: Dissolve 15 g (1 package) of cyanogen bromide-activated Sepharose 4B in 200 ml of 5 mM HCl. Using a large (Millipore) sintered glass funnel and Whatman #50 filter paper, wash the Sepharose with 800 ml of 5 mM HCl, then with 200 ml of water. Dissolve the resin in 100-150 ml of 0.1M $NaHCO_3$-0.3M NaCl ($HCO_3$-NaCl), add 100 mg of poly(U) dissolved in the same buffer (10-20 ml), and swirl gently for two to three hours at room temperature or overnight at $4^o$ C. Do not stir with a magnetic stirrer or the beads will be damaged. After swirling, add 50 ml of 1M propylamine adjusted to pH 8 with HCl. Do this work with adequate ventilation. Swirl another sixty minutes, then wash the resin with 200 ml of $HCO_3$-NaCl and then with 500 ml of water. Suspend in 100-150 ml of 50% EtOH and store at $-20^o$ C. One milliliter of packed resin will bind approximately 250-350 μg of mRNA.

B. Hybridization to Immobilized Poly(U)*

Poly(U) filters or Sepharose columns are washed with binding buffer: 0.1M $NaPO_4$, 0.12M NaCl, 0.1 mM EDTA, 0.01M Tris-HCl (pH 7.3), and 0.5% SDS. The RNA samples are resuspended in the same buffer and applied (a) to columns in approximately 0.1 to 0.3 column volumes of buffer, or (b) to filters in 100-200 μl of buffer (no vacuum). Samples are allowed

*(See Refs. 1, 2 and 5 and Chaps. 1, 6 and 7.)

to bind to filters for five minutes or allowed to flow through columns at approximately 0.5-1.0 ml/min (samples larger than 200 μl are applied to filters at the same flow rate). Filters or columns are washed with binding buffer (no phosphate) to reduce nonspecific binding. Poly(A)-containing RNA is eluted from Sepharose columns with 50% formamide (in 0.05M Tris-HCl, pH 7.5) at room temperature. Poly(U) filters are additionally washed with 0.3M $NH_4Ac$ in 50% EtOH before being counted or before RNA is eluted from them with formamide.

The percentage of poly(A)-containing RNA in a sample is defined as the ratio of poly(U)-binding radioactivity to acid-precipitable radioactivity. Poly(U)-binding data are always corrected for nonspecific adhesion to filters without poly(U).

Depending on the poly(A) content of a given RNA, it may be found that quantitative elution from poly(U) will not occur at room temperature with 50% formamide. In such cases, elevated temperatures (37-45$^{o}$ C) or high formamide concentrations must be used.

It should be noted that poly(U) itself will be eluted from a Sepharose column along with mRNA.

If such mRNA is to be used for cell-free protein synthesis, the poly(U) must be eliminated or the messenger will be highly ineffective in vitro. This may be done by sucrose gradient centrifugation or Sephadex chromatography, after denaturation of poly(A): poly(U) hybrids with 50% formamide.

C. Oligo(dT)-Cellulose Column Chromatography*

Columns of oligo(dT)-cellulose are also used to purify poly(A) or poly(A)-containing RNA. The advantages of oligo(dT)-cellulose are the following: (a) it is an available commercial product and thus need not be synthesized by the investigator; (b) it can be extensively reused; (c) it has a high capacity for RNA (as does poly(U)-Sepharose); and (d) mRNA eluted from it does not contain contaminating polymers (as in the case of poly(U)-Sepharose) that interfere with in vitro protein synthesis or rebinding to other columns with an affinity for poly(A). The principal disadvantages of oligo(dT)-cellulose are: (a) unlike chromatography with immobilized poly(U), there is considerable contamination of mRNA with rRNA. This contamination, however can be eliminated by several cycles through a column; (b) there is a requirement for longer poly(A) chains than are required for affinity to poly(U).

Binding capacities of oligo(dT)-cellulose are detailed by the manufacturer. An appropriate amount of resin is suspended in (dT) binding buffer (0.4M NaCl, 0.01M Tris-HCl, pH 7.5, 1% SDS) and poured into a column,

---

*(See Chaps. 1, 6 and 7.)

usually over a support of glass wool. The column is washed with the same buffer, then with (dT) solution buffer (0.01M Tris-HCl, pH 7.5, 0.05% SDS), and then with (dT) binding buffer again. At this point, RNA is loaded onto the column in binding buffer, at a flow rate of approximately 0.5-1.0 ml/min. Often, all of the mRNA does not bind on the first passage. Therefore, it is recommended that the flow-through be passed through the column again. The column is washed extensively with binding buffer, until no trace of RNA appears in the wash. Bound RNA is step-eluted from the column with (dT) elution buffer. After extensive washing with elution buffer and binding buffer, the column is again ready for use.

### D. Isolation of Poly(A)

Either total RNA or RNA purified by chromatography on oligo(dT)-cellulose or poly(U)-Sepharose can be used as starting material. Poly(A) residues are separated from the bulk of the RNA by virtue of their resistance to the combined action of ribonucleases A and $T_1$ in 2 x SSC (0.30M NaCl, 0.03M sodium citrate) [25]. An ethanol-precipitated RNA sample is resuspended in 0.1-0.3 ml of 2 x SSC containing 10 U/ml of RNase T1 and 5 μg/ml of RNase A (pancreatic RNase). Samples are incubated at $37^{o}$ C for thirty minutes. Nuclease digestion is terminated by the addition of 0.5% SDS and 1 mg of proteinase K.

A subsequent incubation for thirty minutes at 37° C is followed by phenol extraction, and binding to, and elution from, poly(U)-Sepharose. The final poly(A) fragments are then precipitated from ethanol in the presence of 50 μg of E. coli tRNA.

E. Sedimentation Velocity Gradients*

Sucrose gradients are frequently used as a preparative or analytic tool for total RNA or purified mRNA or rRNA. Standard aqueous gradients [1, 2, 5] are 15-30% sucrose in 0.01M Tris-HCl (pH 7.3), 0.1M NaCl, and 0.5% SDS. The two most convenient ultracentrifuge rotors we have used for such gradients are the SW27 and the SW50.1. At 22° C, centrifugation at 49,900 rpm for 2.5 hours in the SW50.1 rotor, or at 24,000 rpm for sixteen to eighteen hours in the SW27 rotor, will sediment 28S rRNA approximately two-thirds of the way to the bottom of the centrifuge tube (the tubes used are either cellulose nitrate or polyallomer). Fractions are collected from the bottom of the tubes,either by puncturing or pumping. If the presence of very large RNA molecules (larger than 45S) is suspected, it is advisable to use a cushion of 50% sucrose in the bottom of the gradient.

Aqueous sucrose gradients will separate RNAs according to size and conformation. Often, aggregates of RNA

*(See Chaps. 1, 5, 6 and 7.)

will give misleading results. To get a more accurate dependence of sedimentation on molecular size, "denaturing" gradients may be used. We have used sucrose gradients containing $Me_2SO$ and also those containing formamide. The $Me_2SO$ gradients contain 0 to 8% sucrose, 99% $Me_2SO$, 1 mM $Na_2$-EDTA, and 0.25 mM TES (pH 6.8) [2]. Formamide gradients are 5 to 20% sucrose, 50% formamide, 0.01M LiCl, 5 mM EDTA, 0.2% SDS, 0.01M Tris (pH 7.4) [26]. Although the formamide gradients are not completely denaturing, they appear to minimize aggregation. They are far easier to run and far more reproducible than $Me_2SO$ gradients.

F. Polyacrylamide Gel Electrophoresis of RNA*

Two polyacrylamide gel systems are used for fractionating RNA: aqueous gels and gels containing 99% formamide.

Aqueous Gels

Gels ranging from 2.4% to 10% acrylamide are prepared in E buffer: 0.04M Tris-HCl (pH 7.3), 0.02M NaAc, 1 mM Na-EDTA (pH 7), and 0.2% SDS [27]. Acrylamide stock solutions are 15% acrylamide, 0.75% bis, E buffer concentrated threefold. Polymerization is initiated by addition of 20 μl of TEMED (tetramethylethylenediamine) and 200 μl of fresh 10% ammonium persulfate/25 ml of gel solution. Flat interphases are obtained by overlaying with water. Lower percent gels, e.g., 2.4 to 3.5%, are diffi-

*(See Chaps. 1, 4, 5, 6 and 7.)

cult to handle and 0.5% agarose is often included to enhance rigidity. Addition of glycerol (10%, v/v) to the gel solution facilitates removal of the gels from the tubes as well as slicing of the gels.

Electrophoresis buffer is the same as E buffer. Sample buffer is also the same, plus 50% glycerol and 0.01% bromphenol blue. Electrophoresis, at five to ten mA/gel, is terminated when the dye band is approximately 4 cm from the bottom of the gel.

Glass and plastic tubes have been used with similar results. The bottoms of the tubes are plugged (for polymerization) with dialysis membranes that are subsequently left on the tubes during electrophoresis.

Formamide Gels

Polyacrylamide gels containing 99% formamide are prepared by the method of Duesberg and Vogt [28]. Formamide is deionized by stirring with 5% (w/v) BioRad mixed-bed resin and buffered with 0.02M phosphate buffer (Na + salts) (pH 7.0). Phosphate salts usually require overnight stirring to dissolve in formamide. The standard gel used is a 3.75% gel and is prepared by dissolving 6 g of acrylamide and 1 g of bis-acrylamide in 165 ml of phosphate-buffered formamide. Polymerization is initiated by addition of 324 μl of TEMED and 2.02 ml of fresh 10% ammonium persulfate. We routinely use the

E. C. Apparatus slab gel box for formamide gels, hence the large volumes described. Gel solutions are poured into the box and allowed to polymerize for at least three hours. Often, to facilitate visualization of slots in the slab gel, a few grains of methyl orange are stirred into the gel solution. After removal of the slot former (when a slab gel is used), slots are overlaid with phosphate-buffered formamide containing 0.01% bromphenol blue. Ethanol precipitates of RNA are dissolved in a solution containing 30 to 50% formamide buffered with 2 mM phosphate, 50 to 70% glycerol and 0.05% cyanol blue ff, heated to 45$^{o}$ C for five minutes, and layered on a gel or gel slot, beneath the formamide "column."

Electrophoresis buffer is 0.04M $Na_2HPO_4$ (pH 7.0). Electrophoresis is at constant voltage (80-100V), usually overnight. In a typical run, the cyanol blue ff marker will migrate 12-15 cm.

Formamide gels are routinely used for separation of mRNA species. The 3.75% gels are excellent for separating bulk mRNA species. For particular species of mRNA, it is often advantageous to go to higher percentage gels. The use of slab gels has permitted us to identify many discrete mRNA "bands" on the gels [29]. To date, attempts to visualize such bands by longitudinal slicing of cylindrical gels have not been successful.

Formamide gels, and in particular those formed in slabs, are preferred for all gel work because: (a) aggregation of RNA species is minimized and (b) diffusion of RNA species in a wet gel is negligible, i.e., during prolonged exposure for preparation of autoradiographs.

G. Fractionation of Polyacrylamide Gels

Slicing

A whole cylindrical gel or a particular column of a slab gel is frozen by being placed on a flat piece of dry ice. Gels are removed from dry ice, fondled until slightly flaccid, and then sliced with a device consisting of parallel razor blades, spaced by washers, and held together with long screws. For counting radioactive material, each gel slice is transferred to a scintillation vial containing scintillation fluid and 3.5% Protosol or 3.5% NCS solubilizer (Caution. These solutions give severe burns). Vials are recapped and shaken overnight at $37^{\circ}$ C before counting.

Elution

Elution is generally used to recover specific mRNA species from formamide-acrylamide slab gels, but it works just as well with other gel systems. In the case of a slab gel, the region of interest is localized by autoradiography of a gel containing $^{32}P$ -labeled mRNA. To properly orient the autoradiogram on the gel, outer edges of the gel are marked with pieces of tape containing discrete marks in radioactive ($^{35}S$ or $^{14}C$) ink. Thus, the gel is covered with Saran Wrap, radioactive ink "markers" are taped on top of the Saran Wrap at the edges of the gel, and x-ray film is placed above all this and exposed. To recover a specific region from the gel, that region is excised from the x-ray film, and the film is then placed back on top of the gel. The film is oriented by virtue of the radioactive ink markers. Subsequently, the hole in the x-ray film serves as a template for dissecting out the appropriate gel region.

To elute the RNA from a specific piece of gel, add to it two volumes of gel elution buffer (0.5M NaAc, 0.01M Tris-HCl (pH 7.4), 0.1 mM EDTA, 1% SDS) and vortex extensively. Then pour this slurry into the band of a disposable syringe with a #18-22 needle. Pass through the needle several times, then centrifuge two to four minutes in an Eppendorf microcentrifuge. Save the supernatant, and repeat the extraction on the pellet. After two such extractions, more than 80 to 90% of the RNA will be eluted. Direct ethanol precipitation of such eluted RNA is unsatisfactory because non-RNA material is also precipitated. This material substantially hinders any further manipulations of such ethanol precipitates. To avoid these problems, eluted mRNA is passed directly over a column of oligo (dT)-cellulose, washed in gel elution buffer, and then eluted from the column with 0.01M Tris-HCl (pH 7.4), 0.01 mM EDTA, and 0.05% SDS. Material purified in this manner can then be ethanol-precipitated without complications. Naturally, this procedure only works for RNAs containing poly(A). To recover eluted, non-adenylated RNA, dilute the eluted sample fivefold with water and add 0.5 volume of ethanol. Pass this through a Whatman CF11 column previously equilibrated with 0.1M NaAc and 33% EtOH [30]. Wash with the same solution, then with 90% EtOH, then with water to elute bound RNA. Cloudy eluates can be filtered. All eluates are subsequently precipitated with ethanol.

Preparative Electrophoresis

Physical extraction of samples from a gel, as above, can often be time-consuming. This problem can be circumvented if one collects samples during electrophoresis from the bottom of a gel. We have previously described a simple and inexpensive procedure for preparative polyacrylamide gel electrophoresis of RNA [31]. The reader is referred to that paper for more details. Basically, our preparative gel is a Plexiglas tube, 0.6 cm (I.D.) and 12.0 cm long.

Near the bottom of the tube, two holes have been drilled, 180$^{o}$ apart. The first, (outlet) hole is approximately 1.9 mm wide and is just large enough to accommodate plastic tubing of the same outside diameter. The second (inlet) hole is merely a pinhole and is sufficiently large to permit a reasonable flow of buffer from inlet to outlet. The outlet tubing has been inserted just far enough into the gel tube to prevent the gel from sliding down to the level of the holes. The lower gel plug has been pushed up against the lower edge of the outlet tubing. This leaves a gap of approximately one to two mm between the two pieces of the gel. Both faces of this gap are parallel.

The outlet tube is connected to a peristaltic pump and hence to a fraction collector. During electrophoresis (in a standard apparatus), buffer is continually pumped from the lower buffer chamber through the inlet hole, across the gap, and out through the outlet hole and tubing. In this way, material migrating out of the gel at any given instant enters the flowing buffer and is transported to a fraction collector. A proper flow rate, determined by electrophoresing color dyes, is adjusted so that no dye enters the lower gel or escapes through the inlet hole. All dye must be transferred to the outlet tube. Under our conditions of electrophoresis, a flow of 0.5 ml/min was found to be sufficient. It should be noted that, for extended runs, the supply of buffer in the lower chamber must occasionally be replenished. For convenience, one can merely attach a siphon from a buffer reservoir to the lower chamber.

Samples eluted in such a manner can be ethanol precipitated directly. Thus, continuous elution eliminates the need to repurify samples on oligo(dT)-cellulose columns or CF11 columns. But, unless one is is equipped with several sets of gel equipment and several fraction collectors, only a small number of samples can be processed in parallel.

### Electrophoretic Elution of Individual Samples

This method combines some of the attributes of methods 2 and 3 (above). A specific gel section is stuffed into the bottom half of a sterile disposable 5-ml pipette plugged with glass wool. A dialysis bag is filled with the appropriate electrophoresis buffer and the pipette containing the RNA sample is half-immersed in the bag. The bag is secured tightly around the pipette and the whole combination is mounted in a standard tube gel apparatus. Upon electrophoresis, the sample is eluted from the gel and into the dialysis bag.

### H. Complete Digestion of RNA for Base Composition Analysis

The RNA samples are digested to completion by incubating in ten to 20 μl of 0.2N NaOH at $37^\circ$ C for sixteen hours or by digestion for two hours at $37^\circ$ C with ten to twenty μl of 2 U/ml RNase T2, 50 μg/ml RNase A, and 50 μg/ml RNase T1, in 0.05M $NH_4Ac$ (pH 4.5). The latter method is preferred, especially when recovery of di-, tri-, and tetraphosphates is important.

## VIII. FRACTIONATION OF DNA

### A. Cesium Chloride and Cesium Sulfate Gradients

Gradients of cesium chloride and cesium sulfate have been used to fractionate Dictyostelium DNA [6, 11] . Use of these gradients for Dictyostelium DNA follows conventional procedures used for all DNAs [33].

### B. Alkaline Sucrose Gradients

We routinely used 5-20% sucrose gradients in 0.3M NaOH, 0.7M NaCl, and 5 mM EDTA (pH 12.6) [34]. Centrifugation in the SW 50.1 rotor is for sixteen hours at 45,000 rpm and $4^\circ$ C. Sucrose solutions should be prepared immediately before use.

C. Agarose Gel Electrophoresis

Dictyostelium DNA, or fragments thereof, can be fractionated on 0.5 to 1.5% agarose gels [35,35]. Agarose is suspended in Tris-borate buffer (10.6 g Tris base, 5.5 g boric acid, 0.93 g Na-EDTA in one liter of $H_2O$, and melted in a boiling-water bath. After melting, ethidium bromide is added to 0.5 μg/ml, and gels are poured in cylindrical tubes, without overlaying. After the gels have hardened, the tops are extruded and cut flat with a razor blade. The DNA is dissolved in Tris-borate + 50% glycerol + 0.1% bromphenol blue + 0.5 μg/ml ethidium bromide. Electrophoresis buffer is the same solution, without glycerol and bromphenol blue. Electrophoresis is at 30 V overnight or 200 V for one to three hours. The DNA in the gel is visualized by illuminating the gel with short- or long-wave UV light.

D. Isolation of Repetitive and Single-Copy DNA

_Dictyostelium_ nuclear DNA, like that of other eukaryotes, contains DNA sequences present in one copy per genome (unique or single-copy DNA), as well as those present in more than one copy per genome (repetitive DNA). Firtel has detailed methods for separating these components and the reader is referred to his paper [11].

### E. Isolation of Poly(dT)$_{25}$ Sequences From Dictyostelium DNA

Recent experiments have shown that Dictyostelium nuclear DNA contains approximately 15,000 tracts of poly(dT)$_{25}$ [6]. These are isolated by chemical depurination of DNA [37], followed by chromatography on poly(A)-Sepharose. Poly(A)-Sepharose is prepared as described above for poly(U)-Sepharose.

## IX. SYNTHESIS OF mRNA AND rRNA PRECURSORS IN ISOLATED NUCLEI [5]

Nuclei are purified as described in Section IV and stored at -70$^{o}$ C in storage buffer. Standard reaction mixtures contain, in a final volume of 100 μl: 0.04M Tris-HCl (pH 7.9), 0.01M $MgCl_2$, 0.01-0.25M KCl, 0.16 mM ATP, GTP, CTP, and UTP, 0.1 mM dithiothreitol, 5% (v/v) glycerol, and nuclei from 5 x $10^5$ to 5 x $10^6$ cells. The reactions are started by the addition of nuclei to the other components and are incubated at 22-23$^{o}$ C for five to twenty minutes; $^{14}$C-, $^{3}$H- or $^{32}$P-labeled ribonucleoside triphosphates are used to monitor transcription. Reactions are terminated by the addition of 25 volumes of 5% (w/v) TCA containing 0.01M sodium pyrophosphate or by the addition of SDS to a final concentration of 0.4%.

Acid-precipitable radioactivity is determined by filtering through Millipore nitrocellulose membrane filters. All data are corrected for background points taken at zero time.

The RNA synthesized in isolated nuclei is subsequently treated in the same manner as the RNA labeled in whole cells with respect to all the fractionation procedures described here.

For extensive synthesis, higher (0.2 to 0.3M) salt concentrations should be used. Also, if insoluble potassium dodecyl sulfate precipitates are a problem, then NaCl is a satisfactory substitute for KCl in all reactions.

## IX. ROUTINE PROCEDURES

### A. Sterility

All glassware and all solutions used are sterilized whenever possible. Where $^{32}$P-labeled material is used, sterile disposable pipettes and beakers are extensively used. Solutions are sterilized by autoclaving or by shaking with a few drops of diethylpyrocarbonate (DEP) followed by boiling to get rid of the DEP. Rubber gloves are used to prevent contact between RNA and putative finger-bound ribonuclease.

### B. Cerenkov Counting

Whenever $^{32}$P-labeled material is used, we spare ourselves the cost of scintillation fluid by counting samples directly in the $^{3}$H channel of a scintillation counter. Such Cerenkov counting is approximately 30% as efficient as counting with scintillant, but the samples may be recovered.

The use of $^{32}$P-labeled material also permits all stages of an experiment to be monitored with a hand-held radiation monitor. We have found that those monitors emitting an audible signal are particularly useful.

### C. Siliconizing Glassware

In situations where very small amounts of RNA or DNA are handled, significant losses can be incurred by adhesion of solutions to glass. For this reason, glassware that is to contain less than 50 μg of nucleic acid is usually siliconized. This is done by rinsing glassware in 5% dichloro-dimethyl silane in benzene (use in a hood--deadly fumes), air drying, rinsing in distilled water, and baking at 350$^{\circ}$ for one to five hours.

## X. SOURCES OF REAGENTS

Proteose peptone, yeast extract,
  Difco Labs, Detroit, Mich.
All radioisotopes, Protosol,
  New England Nuclear, Boston, Ma.
Cemusol NPT-12,
  Nelle-Bezons, 92-Neiully-sur-Seine, France
SDS,
  BDH Chemicals, Poole, England
Diethylpyrocarbonate,
  Eastman Organic Chemicals, Rochester, N.Y. or
  Calbiochem, La Jolla, Calif.
Proteinase K,
  E. M. Labs, Kankakee, Ill. or Sigma, St.
  Louis, Mo.
Cyanogen bromide-activated Sepharose 4B,
  Pharmacia Fine Chemicals, Piscataway, N.J.
Oligo(dT)-cellulose, Type T2 or T3,
  Collaborative Research, Waltham, Ma.
E. C. slab gel box,
  E. C. Apparatus Co., St. Petersburg, Fla.
Formamide, spectral grade,
  Matheson, Coleman and Bell, E. Rutherford, N.J.
Dichloro-dimethyl silane,
  Aldrich Chemical Co., Milwaukee, Wis.
Nucleotides, Nucleosides,
  P. L. Labs, Milwaukee, Wis.
RNase A,
  Worthington Biochemicals, Freehold, N. J.
RNases T-1 and T-2,
  Calbiochem, La Jolla, Calif.
Acrylamide and bis-acrylamide,
  Bio Rad Labs, Richmond, Calif.
Cesium chloride and cesium sulfate (special
  biochemical grade),
  Gallard-Schlesinger Chem. Mfg., Carle Place, N.Y.
Agarose(electrophoresis grade),
  Sigma, St. Louis, Mo.
Ethidium bromide,
  Calbiochem, La Jolla, Calif.

All other chemicals used are those of the highest purity available from Sigma, Calbiochem, Fisher, S/P etc.

ACKNOWLEDGMENTS

I am indebted to my colleagues, Drs. Harvey Lodish, Charles Lane, Richard Firtel, and Maurice Sussman, all of whom have had a major part in establishing the methods described here.

REFERENCES

1. R. A. Firtel, A. Jacobson, and H. F. Lodish, Nature(New Biol.), 239, 225(1972).
2. R. A. Firtel, and H. F. Lodish, J. Mol. Biol., 79, 315(1973).
3. R. A. Firtel, L. Baxter, and H. F. Lodish, J. Mol. Biol., 79, 315(1973).
4. H. F. Lodish, R. A. Firtel, and A. Jacobson, Cold Spring Harbor Symposium Quant. Biol., 38, 899 (1973).
5. A. Jacobson, R. A. Firtel, and H. F. Lodish, Cold Spring Harbor Symposium Quant. Biol., 38, 899 (1973).
6. A. Jacobson, R. A. Firtel, and H. F. Lodish, Proc. Natl. Acad. Sci. U.S.A.,71, 1607 (1974).
7. H. F. Lodish, A. Jacobson, R. A. Firtel, T. Alton, and J. Tuchman, Proc. Natl. Acad..Sci. U.S.A., 71; 5103 (1974).
8. A. Jacobson, R. Firtel, and H. F. Lodish, Brookhaven Symp. In Press (1974).

9. J. Darnell, W. Jelinek, and G. Molloy, Science, 181 181, 1215 (1973).

10. R. Weinberg, Annu. Rev. Biochem., 42;329 (1973).

11. R. A. Firtel, and J. Bonner, J. Mol. Biol., 66, 339 (1972).

12. D. J. Watts, and J. M. Ashworth, Biochem. J., 119, 171(1970).

13. S. M. Cocucci and M. Sussman, J. Cell. Biol., 45, 399(1970).

14. M. Sussman in Methods in Cell Physiology, Vol. 2,Academic Press, New York, 1966,p. 397.

15. J. Bonner, The Cellular Slime Molds, Princeton University Press, Princeton, 1967.

16. P. C. Newell, Essays in Biochem., 7, 87(1971).

17. M. Sussman, and R. R. Sussman, Symp. Soc. Gen. Microbiol. 19, 403 (1969).

18. M. Sussman, personal communication.

19. F. Rothman, personal communication.

20. R. Kessin, Ph.D. thesis, Brandeis University, 1971.

21. A. Jacobson, and H. F. Lodish, unpublished experments.

22. R. Roberts, personal communication.

23. R. W. Davis, personal communication.

24. R. Sheldon, C. Jurale, and J. Kates, Proc. Natl. Acad. Sci. U.S.A., 69, 1321 (1972).

25. C. W. Anderson, J. B. Lewis, J. F. Atkins, and R. F. Gesteland, Proc. Natl. Acad. Sci. U.S.A., 71, 2756 (1974).

26. C. W. Anderson, J. B. Lewis, J. F. Atkins, and R. F. Gesteland, Proc. Natl. Acad. Sci. U.S.A., 71, 2756 (1974).

27. D. H. L. Bishop, J. R. Claybrook, and S. Spiegelman, J. Mol. Biol., 26, 373 (1967).

28. P. H. Duesberg, and P. K. Vogt, J. Virol., 12, 594 (1973).

29. A. Jacobson, C. Lane, and T. Alton, in preparation (1974).

30. D. Gillespie, personal communication.

31. A. Jacobson, and H. F. Lodish, Analytic Biochem., 54, 513(1973).

32. B. Barrell, in Procedures in Nucleic Acid Research, G. L. Cantoni and D. R. Davis, eds., Harper and Row, New York, 1971, p. 751.

33. W. Szybalski and E. H. Szybalski in Procedures in Nucleic Acid Research, G. L. Cantoni and D. R. Davis, eds., Harper & Row, New York, 1971, p. 311.

34. I. Verma, G. F. Temple, H. Fan, and D. Baltimore, Nature(New Biol.), 235, 163(1972).

35. A. Jacobson, unpublished experiments.

36. R. A. Firtel, personal communication.

37. K. Burton in Methods in Enzymology, L. Grossman and K. Moldave, eds. Vol. 12 A, Academic Press, New York, 1967, p. 222.

Chapter 4

# ANALYSIS OF ERYTHROID DEVELOPMENT

Allen J. Tobin, Hildur V. Colot, Joanne Kao,
Kay S. Pine, Scott Portnoff,
Nicholas N. Zagris, & Nancy Zarin
Biology Department & Molecular Biology Institute,
University of California, Los Angeles, California &
The Biological Laboratories, Harvard University,
Cambridge, Massachusetts

## I. INTRODUCTION

Knowledge of how globin genes are regulated during erythropoiesis and during ontogeny is important both to the understanding of hemapoietic disorders and to the formation of a general picture of eukaryotic regulation. Red cell development provides many advantages for the study of regulation at the molecular level: (a) blood is an easily available tissue, typically totaling 8% of the body weight of a vertebrate; (b) there is enormous specialization in erythrocytes, with hemoglobin comprising more than 95% of their cytoplasmic protein; (c) hemoglobin messenger RNA (mRNA) is easily isolated on the basis of size and of the presence of polyadenylic acid sequences [1-3]; and (d) there are easily detectable changes in the pattern of hemoglobin synthesis during development in all vertebrates that have been studied [4,5].

We have chosen to study erythropoiesis in the chicken for several reasons: (a) chickens and their embryos are relatively inexpensive and available throughout the year; (b) early chick embryos are accessible and manipulable, making possible studies that extend over the entire course of normal erythroid development; (c) it is possible to study chick erythropoiesis either in organ cultures of early blastoderms or in cell cultures derived from 18 to 36 hour embryos [6-8]; (d) chick erythroblast cultures derived from the early embryos make all the hemoglobins found in the course of normal development, whereas the mammalian erythroblasts that have been studied make only adult hemoglobins in culture [8-11]; (e) the presence of large numbers of circulating erythroblasts with active nuclei both in anemic adults and in embryos allows the study of nuclear RNA metabolism in a highly specialized cell type [12-14].

On the other hand, mammalian systems would seem a priori to have greater relevance to the human condition, as well as some important experimental advantages. The availability of hemoglobin variants and genetic hemapoietic disorders in mice make them an especially attractive species for study [15, 16]. In sheep and goats, either anemia, hypoxia, or the transfusion of plasma from an anemic animal can induce the transient synthesis of a new hemoglobin [17, 18]. Since this switch and its reversal can now be effected in cultured sheep erythroblasts, this fascinating system has become an accessible and important object of study [10, 11].

Studies of mammalian erythropoiesis are expensive, however, and are also limited by the lack of active, nucleated erythroblasts in the adult circulation. Such active erythroblasts can be obtained in mammals only from the early embryo or from bone marrow. Biochemical experiments are limited in the former case by the amount of material and in the latter by the heterogeneity of the cell population.

During the course of normal chicken ontogeny, five striking changes occur in the relative amounts of individual hemoglobins:

The first appearance of hemoglobins E, M, P, P', and P", at about 36 hours of incubation.
The first appearance of hemoglobins A and D at about 6 days of incubation.
The disappearance of the early hemoglobins at about 12 days of incubation.
The transient appearance of hemoglobin H between day 12 of incubation and a few days after hatching.
The change in the ratio of hemoglobin A to hemoglobin D from 1:1 at 6 days of incubation to 4:1 in the adult [19].

This chapter describes the methods we are using to study these changes and to elucidate their molecular mechanisms. The bulk of this chapter deals with means of analyzing the relative amounts and rates of synthesis of individual hemoglobins and globin chains. Most of the methods for the analysis of RNA levels and synthetic rates are discussed in detail in this volume by Jacobson (Chap. 3) and by Efstratiadis and Kafatos (Chap. 1), and we therefore describe only those methods that are particularly adapted for the study of erythroblast RNA.

## II. COLLECTION OF BLOOD FROM CHICKENS AND THEIR EMBRYOS

We collect up to 10 ml of blood from the wing veins of adult chickens either into vacutainers (Becton-Dickinson) containing heparin or into 10-ml syringes containing 1 ml of 10 mg/ml sodium heparin in isotonic saline (0.14M NaCl, 0.05M KCl, 1.5 mM $MgCl_2$, hereafter called NKM).

To obtain erythroblasts, we induce anemia in white Leghorn roosters (SPAFAS, Norwich, Connecticut) weighing 1.5 to 2.5 kg by intramuscular injection of 15 mg/ml acetylphenylhydrazine in 0.1M $Na_2HPO_4$ (pH 7.4), according to the following schedule: day 1, 2.0 ml/kg; day 2, 1.5 ml/kg; day 3, 1.0 ml/kg; day 4, 0.5 ml/kg. On the fifth day, roosters can be bled from the wing vein as described above or they may be killed to obtain fifty to one hundred ml of blood from the jugular vein. In the latter case, they are injected intravenously with 500 units/kg of Nembutal and 500 units/kg of sodium heparin, and then bled by cutting one of the jugular veins.

The degree of anemia is characterized by the

reduction in the hematocrit (the fraction of the blood volume occupied by erythrocytes) and by the presence of immature cells in the circulation. With the procedure described above and white Leghorn roosters, the normal hematocrit of 0.40 to 0.45 was reduced to 0.15 to 0.20 with about 20% rooster mortality. To determine the presence of immature erythroid cells, we make blood smears, fix them 5 minutes in methanol, then stain them twenty minutes in 2% Wright's (Harleco), 6% Giemsa (Harleco). We then type them according to the nomenclature of Lucas and Jamroz [20]. Ordinarily, we obtain 1 to 3% basophilic erythroblasts, twenty to thirty% mid-polychromatophilic erythroblasts, and mature erythrocytes.

More than 95% of the cells from an anemic rooster can be classified as reticulocytes on the basis of supravital staining with new methylene blue [21]. This is done by mixing equal volumes of blood and 0.5% new methylene blue in 1.6% potassium oxalate; after this mixture is incubated for fifteen minutes at room temperature, a smear is made, and the fraction of cells containing blue granules (i.e., the reticulocytes) is determined.

We obtain embryonic erythroid cells from white Leghorn embryos of various ages. Fertile, pathogen-free, white Leghorn eggs (SPAFAS, Norwich, Connecticut), stored no more than six days at 16$^{o}$ C, are incubated at 38$^{o}$ C in a humidified incubator. We use a Model 55 Humidaire Incubator (Humidaire, New Madison, Ohio), which turns the eggs through about 60$^{o}$ every hour.

To bleed embryos from sixty hours to six days old, we remove the shell and shell membrane from the blunt end of the egg and break the blood vessels around the heart by sucking the heart into a Pasteur pipette that has been broken and heat polished so that its tip diameter is slightly larger than that of the heart. Blood cells are washed free of solid tissue by transferring the pipette's contents to a Petri dish containing cold NKM. The embryo is allowed to bleed for two to five minutes, and the pooled blood is transferred to a test tube containing cold NKM.

To bleed embryos older than six days we take advantage of the attachment of the embryonic membranes to the shell near the blunt end of the egg. We cut a hole at the pointed end of the egg, cut the yolk sac, and allow the yolk and albumin to drain. Taking care not to break the blood vessels prematurely, we then make a large hole in the shell, wash the embryo free of yolk with cold NKM, cut the major blood vessels with scissors, and collect the blood with a Pasteur pipette into cold NKM. Blood from embryos is filtered through four layers of cheesecloth and then through monofilatment nylon screen (Nitex 45: Tobler, Ernst, and Traber, New York) of a 45-μm pore size to remove contaminating solid tissue.

Cells are collected from NKM and plasma by centrifugation for three minutes at full speed (2000 g) in a clinical centrifuge. The cells are then washed three times in cold NKM (3 to 25 volumes) and lysed by mixing with three volumes of a lysis solution containing 0.03M KCl, 0.02M $MgCl_2$, and 0.01M Tris-HCl (pH 7.4), to which has been added 0.1% Triton X-100. After stirring the lysis mixture five minutes in ice, we remove nuclei and cell membranes either by centrifugation for ten minutes at 27,000 g at $4^\circ$ C or by mixing the lysate with 0.5 volume of one part toluene to one part carbon tetrachloride, centrifuging ten minutes at 2000 g, and recovering the hemoglobin layer. An aliquot of the supernatant is used to determine the total hemoglobin concentration and the rest of the supernatant is either analyzed immediately or frozen in liquid nitrogen.

We determined the total hemoglobin concentration in a lysate after conversion to cyanmethemoglobin by dilution in fifty volumes of Drabkin's solution, 0.1% $NaHCO_3$, 0.005% KCN, and 0.02% potassium ferricyanide [22]. The absorbance of this solution is measured against Drabkin's solution diluted with 0.02 volume of Tris-HCl. The hemoglobin concentration in milligrams per milliliter of hemoglobin in the diluted solution is given by $A_{420} \times 0.1605$ or by $A_{540} \times 1.54$ [19].

## III. PREPARATION OF HEMOGLOBINS FOR ANALYSIS

The most commonly used method for the analytic separation of hemoglobins is electrophoresis, with starch gel, cellulose acetate, agar, or polyacrylamide gel as a supporting medium. Since the mammalian and avian hemoglobins that have been studied have the same molecular weight and presumably the same shape, their electrophoretic separation depends only on differences in charge. Since the charge of a protein is a function of the pH and ionic strength, the ability to resolve two hemoglobins or more by electrophoresis is a function of ionic conditions. These must be empirically determined for a given set of hemoglobins.

Insufficient charge differences at the running pH can obviously result in more than one hemoglobin in a given electrophoretic component. Conversely, a single hemoglobin can give more than one electrophoretic band as a result of the oxidation of some or all of the ferrous ions in the four heme groups. To avoid the latter difficulty, it is necessary to protect the hemoglobins from oxidation by saturating the hemes with carbon monoxide or cyanide or to convert all the hemoglobin to the same oxidized form, usually cyanmethemoglobin. To convert hemoglobins to carbonmonoxyhemoglobins, we pass carbon monoxide from a fifty ml syringe over a lysate. Carbonmonoxyhemoglobins are stable for many weeks in the cold, but we generally use these lysates within a day. We obtain cyanmethemoglobin by mixing an aliquot of lysate with an equal volume of a solution containing 0.5% KCN, 2% potassium ferricyanide, and 0.1% sodium bicarbonate.

## IV. ANALYSIS OF HEMOGLOBINS BY ELECTROPHORESIS

Electrophoretic analysis of chicken hemoglobins is performed by the method described by Bruns & Ingram [19].

Stock solutions for electrophoresis are prepared from the following components, brought to 100 ml with glass-distilled water:

| Stock Solution | Components/100 ml of Solution | |
|---|---|---|
| A | 30 g | of acrylamide (electrophoresis grade or recrystallized from chloroform) |
| | 0.8 g | of N, N'-methylene-bis-acrylamide (bis) (electrophoresis grade or recrystallized from chloroform) |
| B | 0.33 g | of N, N, N', N'-tetramethylethylene diamine (TEMED) |
| | 24 ml | of 1N HCl |
| | 5 g | of Tris (enzyme grade, pH 7.9) |
| C | 0.14 g | of $(NH_4)_2S_2O_8$;(freshly made) |
| D | 5 g | of acrylamide |
| | 1.25 g | of bis |
| E | 0.1 ml | of TEMED |
| | 12.8 ml | of 1M $H_3PO_4$ |
| | 2.85 g | of Tris(pH 7.2) |
| F | 0.28 g | of $(NH_4)_2S_2O_8$ |
| F' | 2 mg | of riboflavin |

To analyze ten samples by electrophoresis in cylindrical polyacrylamide gels, we prepare 20 ml of resolving gel solution by mixing 5 ml of A, 5 ml of B, and 10 ml of C. This solution is degassed by suction and poured into glass tubing (i.d. 5 mm) to a height of 6 cm. Water (about 50 μl) is layered onto the top of the gel solution in each tube to give a flat gel surface. Polymerization occurs in about thirty minutes at room temperature.

The stacking gel solution is prepared by mixing 2 ml of D, 1 ml of E, and 1 ml of F or F' and degassing. The water is removed from the top of the polymerized resolving gels, and the surface of each gel is rinsed with a few drops of the stacking gel solution, which are then removed and discarded. The stacking gel solution is then added to a height of 1 cm in each tube, and 50 μl of water is layered on top before polymerization. When photopolymerization is used, stock F' is substituted for F and the gels are exposed to a fluorescent light.

After polymerization the gels are immediately put into a vertical cylindrical gel electrophoresis apparatus, and the upper (cathode) and lower (anode) reservoirs are filled with electrode buffer, which consists of 6.32 g/liter of Tris and 3.94 g/liter of glycine (pH 8.9). When hemoglobins are separated as their cyanmet derivatives, 0.1 g/liter of KCN is added to the electrode buffer. Samples containing 10 - 100 μl are mixed with 0.5 volume of 50% glycerol. The glycerol increases the density of the sample so that it can be layered beneath the electrode buffer, on top of the appropriate gel.

Electrophoresis is performed at $4^{\circ}$ C, with the gels run at 1 mA/tube for twenty minutes and 2 mA/tube for 150 minutes. After they are run, we remove the gels from their tubes by injecting water between the gel and the glass wall with a long syringe needle and forcing the gel out of the tube with gentle air pressure. The gels are then put either into water and scanned immediately at 540 nm or into a fixative, generally 10% TCA. The unstained gels can be photographed through a green filter.

If the gels are to be stained, after two hours in 10% TCA, they are washed one hour in 7.5% HAc and stained overnight in 0.006% Coomassie Brilliant Blue in a mixture of five volumes of MeOH, five volumes of $H_2O$ and one volume of HAc. The gels are destained in a mixture containing 7.5% HAc and 5% MeOH, and stored in 7.5% HAc. The stained gels can be scanned at 590 nm or photographed through an orange filter.

For complex mixtures of hemoglobins, such as those found in many developing vertebrates, including chickens, isoelectric focusing gives resolution superior to electrophoresis [23]. The ionic conditions that have been optimized for the electrophoretic separation of components A and B are usually suboptimal for the separation of components C and D, whereas the establishment of a linear pH gradient in isoelectric focusing assures the resolution of any proteins whose isoelectric points (pI) differ by more than 0.03 pH units (e.g., human hemoglobins A, with pI = 6.95, and S, with pI = 7.25, are easily resolved). Figure 1 compares the separation of chicken hemoglobins by isoelectric focusing and by electrophoresis in cylindrical polyacrylamide gels [24].

To analyze hemoglobins by isoelectric focusing, we prepare the following stock solutions:

| Stock Solution | Components/100 ml of Solution |
|---|---|
| A (same as A above) | 30 g of acrylamide<br>0.8 g of Bis |
| C (same as C above) | 0.14 g of ammonium persulfate, freshly made |
| G | Ampholine (pH 7-9), 40% stored at $4^{o}$ or - $20^{o}$ C |

To analyze ten samples, we prepare 20 ml of solution by mixing 3.33 ml of A, 10 ml of B, and 5.67 ml of glass-distilled water. This solution is degassed for five to ten minutes, and 1 ml of G is added. No TEMED is needed,since the ampholines themselves catalyze the polymerization of the acrylamide. The final mixture is pipetted into cylindrical glass tubing (5mm, i.d.) to a height of 7 cm. Water (about 50 μl) is layered on top to give a flat surface.

After the gel polymerizes, the water layer is removed and 50 μl of 10% glycerol in Tris is added. This layer protects the sample from the high pH of the cathode reservoir. The hemoglobin sample, converted to its cyanmet form as described above, is mixed with one volume of 40% sucrose in water and loaded under the protective layer. The gel tubes are then placed into a vertical cylindrical gel electrophoresis apparatus. The tubes and the upper (cathode) reservoir are filled with 2% ethylene diamine. The lower (anode) reservoir is filled with 0.2% sulfuric acid. After ten minutes at 50 volts and ten minutes at 100 volts, the potential across the gel is in-reased to 200 volts. After three to four hours at 200 volts, the hemoglobin bands are sharply focused. The gels are then removed from the tubes, fixed, and stained as above.

## V. SEPARATION OF HEMOGLOBINS ON POLYACRYLAMIDE GEL SLABS

Both electrophoresis and isoelectric focusing can be performed on slab gels [25]. Slab gels have several advantages over cylindrical gels: (a) they allow a more convenient comparison of samples; (b) since the surface area of a slab is greater than that of a cylinder of the same cross-sectional area, more power can be dissipated, and separation can be accomplished more rapidly; (c) the use of thin slabs requires less gel solution than standard cylindrical gels, an especially important financial consideration with isoelectric focusing; (d) the amount of hemoglobin needed per sample can be much lower than in the case of cylindrical gels; using slab gels we routinely detect as little as 2 to 3 μg of a single hemoglobin component after staining with Coomassie Brilliant Blue, whereas about 10 μg is required in 5-mm cylindrical gels; (e) the slabs are conveniently dried for storage or exposure to X-ray film for autoradiographic or fluorographic determination of labeled products.

Vertical slab gel apparatuses are commercially available or they may be built inexpensively. For polyacrylamide gel electrophoresis we use a Plexiglas vertical apparatus with a machined cooling plate through which tap water is circulated. We usually perform electrophoresis on slabs without a stacking gel. Vertical slabs are cast upside down, with a template for the sample wells on the bottom. If a stacking gel were used, it would be cast first and the running gel poured on top. As long as sample volumes are small ($\leq$ 100 μl), however, we have found that a stacking gel is not necessary for excellent resolution of the normal chicken hemoglobins, either in slabs or in cylinders.

For isoelectric focusing in an acrylamide slab gel, we use a horizontal apparatus whose design is shown in Figure 2. We prepare the following stock solutions and pass them through 0.45-μm Millipore filters to remove insoluble impurities.

| Solution | Components/100 ml of Solution |
|---|---|
| A (same as A above) | 30 g of acrylamide<br>0.8 g of Bis |
| G (same as G above) | Ampholine (pH 7-9), 40% |
| H | 5% amonium persulfate, freshly made |

Two 13 x 11.5 cm glass plates and one 8.7 x 11.5 cm plate are washed carefully in a strong detergent, then rinsed with water, 60% EtOH, and distilled water. The plates are drained and wiped dry with Kimwipes or lens paper; care must be taken to avoid fingerprints on the plates. A 1/32" (0.8 mm) rubber gasket that fits the perimeter of the larger plates is also washed and wiped dry. With gloved hands, we then clamp together with metal paper clamps (IDL binder clips), the two larger plates and the gasket. The top edge of the gasket has an opening through which the gel solution is inserted. This assembly is held upright by clamps used as supports. We have also used a 1/16" (1.5 mm) gasket. The resulting thicker gels are easier to handle but require twice the material and take longer to uun.

Into a 50-ml suction flask, we put 7.7 ml of filtered distilled water, 1.67 ml of A, and 0.5 ml of G. This mixture is then evacuated with a water pump and swirled frequently to facilitate the removal of dissolved gas. When bubbles stop forming we add 0.13 ml of H and then immediately add the completed gel solution to the assembled apparatus with a Pasteur Pipette or a syringe. This must be done quickly since polymerization is rapid. If the assembly is not to be completely filled with gel solution, we layer water on top of the gel solution and tilt the assembly back and forth until there is a level interphase between the water and the gel solution. Polymerization is complete in less than twenty minutes at room temperature. The gel can be used immediately or stored at $4^{o}$ C for up to ten days, if care is taken that the gel does not dry out.

When the gel is to be used, the glass plates of the mold are gently pried apart, and the upper surface of the gel is wet with 10% glycerol. The smaller plate is then placed on top of the gel so that there is a 2-cm margin of gel extending beyond the top plate on each side. The top plate is put into place like a microscope cover slip, starting at one edge and forcing the liquid and bubbles ahead of it as it is laid down. This liquid is removed with a pipette and by blotting.

Two stacks of three 1/2" (1.25 cm) strips of Whatman 3MM filter paper are put at the extreme ends of the gel to serve as reservoirs of the electrode solutions. The strip in contact with the cathode is soaked in 1% ampholine (pH 9-11), and the strip in contact with the anode with 0.1% ampholine (pH 7-9). The samples are converted to cyanmethemoglobins or carbonmonoxyhemoglobins. Small pieces (2 x 5 mm) of Whatman GF/A glass-fiber filters are then soaked in the sample solutions, gently blotted, and placed on the anodic side of the gel, as close as possible to the top plate. Greater volumes may be loaded onto stacks of three filters. The gel assembly is then put on top of a cooling plate and the voltage acrsos the gel is increased to 800 V during a 10 to 15 minute period. We generally do this by setting our power supply to a constant current of 10 mA and allowing the voltage to climb to 800 V as the conductance of the gel decreases. We then switch the power supply to a constant voltage of 800 V, which we maintain for two hours. The current may be briefly interrupted and restarted without affecting the separation.

After focusing is complete, the gel is removed from the plates and fixed in 10% TCA for at least one hour. We usually stain our gels by washing them thirty minutes in 7.5% HAc, staining them one to two hours in 0.006% Coomassie brilliant blue in MeOH: $H_2O$: HAc (5:5:1) and destaining them in 7.5% HAc, 5% MeOH for at least one hour.

The slab gels may be dried for convenient storage or for autoradiography or fluorography [25]. The gel is placed on a piece of Saran Wrap, and the wrinkles are smoothed out. A piece of Whatman # 1 filter paper is carefully placed on top, and this sandwich is put onto a thin rubber sheet (50 cm x 60 cm x 0.25 mm) and covered with a piece of 3/16" (0.5 cm) porous linear polyethylene (BoLab, Keene, N.H.). A second rubber sheet is now placed on top; this top sheet has a hole cut in its center through which a polyethylene connector is inserted for connection to a vacuum. This assembly is placed on top of a 8 x 12" (20 x 30 cm) metal baking pan above a boiling-water bath. A vacuum is provided by a vacuum pump through a trap in dry ice and methanol. When the vacuum is applied, the edges of the two rubber sheets seal and the liquid in the gel is removed in one to two hours. After the gel is dry, the Saran Wrap may be easily removed and the dried gel on filter paper stored in a notebook. Some of the steps in the preparation of the gel described above are indicated in Figure 3.

## VI. ANALYSIS OF GLOBIN CHAINS

The globin chain composition of cell lysates, isolated hemoglobins, or the products of cell-free synthesis may be analyzed by isoelectric focusing in 6M urea. Before such analysis, it is necessary to remove heme, the presence of which decreases the dissociation into individual globin chains and can cause the spurious aggregation of denatured protein, and to block the free sulfhydryl groups of cysteinyl residues, which otherwise will form inter- and intramolecular disulfide bonds and produce artifactual heterogeneity.

Heme is removed and protein is precipitated in cold acid-acetone [26]. This solution is made by mixing 100 ml of cold ($-20^{\circ}$ C) acetone with 5 ml of glacial acetic acid, 2 ml of concentrated hydrochloric acid and 0.1 ml of mercaptoacetic acid. A sample containing 10-100 μg of hemoglobin is slowly added to

10 volumes (or a minimum of 0.5 ml) of acid-acetone, and the mixture is allowed to sit in ice for five minutes. The precipitated protein is collected by filtration onto Whatman GF/C glass-fiber filters. We collect small samples by filtration through a "micro-chimney," a piece of 3 mm (i.d.) glass tubing, the ends of which have been polished.

The precipitate is washed in place twice with cold acid-acetone and twice with cold acetone. The samll spot of precipitate on the filter is then cut out and placed in 150 μl of freshly made urea buffer. This buffer consists of 8M urea, 5 mM EDTA, 0.58M Tris-HCl (pH 8.6 to 8.75); it is made by combining 0.72 g of urea, 0.6 ml of 1.44M Tris base dissolved in 0.3N HCl, 0.06 ml of 0.134M $Na_2$ EDTA, and bringing the total volume to 1.5 ml with distilled water. Sulfhydryl groups are then blocked with ethyleneimine [27]. The test tubes containing samples are flushed with nitrogen, and 2 μl of 2-mercaptoethanol is added to each tube. After thirty to sixty minutes at room temperature, 5 μl of ethyleneimine is added. After 120 minutes, another 2 μl of ethyleneimine is added, and the aminoethylation is continued for another ninety minutes. The samples are then frozen in liquid nitrogen for subsequent analysis by isoelectric focusing.

Analysis of globin chains is performed in 5% acrylamide cylindrical gels containing 6M urea [28]. The following stock solutions are prepared:

| Solution | Components /100 ml of Solution |
|---|---|
| I | 10 g of acrylamide<br>0.4 g of Bis |
| J | 0.56 g of ammonium persulfate, freshly made |
| G | Ampholine (LKB) (pH 7-9), 40% |
| K | Ampholine (LKB) (pH 9-11), 20% |
| | **Components /5 ml of Solution** |
| L | 0.9 g of urea<br>0.5 ml of 1.44 M Tris in 0.3M HCl |

To prepare six cylindrical gels, we put into a 50-ml suction flask 3.6 g of urea, 5 ml of I, and 0.627 ml of J. This mixture is brought to 9.25 ml with distilled water and degassed. We then add 0.25 ml of G and 0.50 ml of K, mix, and pour the resulting solution into the gel tubes. About 50 μl of water is layered on top to give a flat surface. Polymerization is complete within one hour. The tubes are mounted in a vertical gel electrophoresis apparatus and filled with 2% ethylenediamine. We put 50 μl of L on top of each gel as a protective layer and apply the sample under it. The upper (cathode) reservoir is filled with 2% ethylenediamine, and the lower (anode) reservoir is filled with 0.2% $H_2SO_4$.

Focusing is done at $4^o$ C. The potential across the gels is raised to 200 V after ten minutes (each) at 50 V and 100 V. After eight to twelve hours, the gels are removed, fixed, stained, scanned, sliced, and counted as described above.

## VII. IN VITRO HEMOGLOBIN SYNTHESIS

To study the relative rates of synthesis of individual hemoglobins and globin chains, we incubate erythroblasts from anemic adults or from embryos of various ages in a medium containing a labeled amino acid, generally [$^3$H]leucine. Cells are obtained and washed as described above, except that embryonic erythroblasts are washed with NKM alone. After three such washes, the cells are washed once with ten volumes of that medium. We routinely use a medium consisting of Dulbecco's modified Eagle's Medium supplemented with antibiotics, 2 mM fresh glutamine, 10% dialyzed fetal calf serum (GIBCO), 0.8 μg/ml of $FeSO_4$, 0.9 μg/ml of ZnCl, and 2 ng/ml of $CuSO_4$. When we wish to label hemoglobins to higher specific activities than are obtained in this medium, we use leucine-free Minimal Essential Medium, supplemented with 10% dialyzed fetal calf serum. The cell suspensions are placed in 10-cm Petri dishes (5 to 15 ml in each) or in 75 $cm^2$ tissue culture flasks (40-100 ml in each). The cells are incubated at 37° C in a 5% $CO_2$ atmosphere and are kept in suspension by swirling them at 60 rpm on a rotating table inside the carbon dioxide incubator. After a thirty minute incubation, to allow thermal and ionic equilibration, labeled amino acid is added. We use 10 to 100 μCi of [$^3$H]leucine (10-20 Ci/mmole) per milliliter of the cell suspension.

Total incorporation of labeled amino acid into acid-insoluble material is monitored as a function of time by spotting aliquots up to 50 μl onto 2.4 cm discs of Whatman #3MM paper. These discs are numbered in pencil before the start of an experiment. After the incubation is completed, the discs are dried and stacked. A 10-cm piece of cotton thread is pulled through the stack with a large sewing needle, and the ends are tied together. The discs are now processed together by being placed successively into 100-200 ml of the following:

1. 10% TCA, 3% casamino acids, at room temperature, 10 minutes

2. 5% TCA, 3% casamino acids, 100°C, 5 minutes. Theyare rinsed successively in 100 ml of
3. 5% TCA, 3% casamino acids, at room temperature
4. 5% HAc, 2% HCl in acetone
5. 95% EtOH
6. 100% EtOH
7. 3 parts 100% EtOH, one part ether
8. Ether

They are then dried in air and counted in a toluene-based scintillation cocktail.

Using these methods of culture and analysis,incorporation of [$^3$H]leucine into material insoluble in hot TCA, i.e., into protein, is linear with time for at least one hour.

To determine incorporation into individual hemoglobins, the incubation is stopped by the addition of 1 mM unlabeled leucine to the cell suspension and the cells are immediately spun away from the medium at 2000 g for three minutes. The cells are washed once in ice-cold medium, and then lysed and analyzed as described above. After electrophoresis or isoelectric focusing is completed, the gels are washed for 24 to 48 hours in several changes of cold 10% TCA to remove unincorporated leucine. We fix cylindrical gels by placing them in numbered plastic tubes (12 mm diameter), which have been perforated with a hot needle. These tubes are placed in a beaker containing 1000 ml of 10% TCA and stirred in the cold.

Slab gels are put into a large Pyrex baking dish containing 500 ml of 10% TCA. This is especially important in the case of isoelectric focusing, since free leucine (pI = 5.98) will focus within the gels we use. After fixing the proteins and washing away small molecules, we scan cylindrical gels at 540 nm to determine the positions and relative amounts of the hemoglobins. We measure the label in individual components in cylindrical gels by freezing the gels in powdered dry ice and cutting them into 1 mm slices with an array of razor blades and washers. The gel slices are placed in small scintillation vials and incubated at room temperature for 24 hours in 4 ml of 6% Protosol (NEN) and 0.4% Omnifluor (NEN) in toluene. Radioactivity is then determined in a liquid scintillation counter.

Radioactivity in the individual components isolated in acrylamide gel slabs is measured by the fluorographic method of Bonner and Laskey [29]. This is more economical of time and materials than the slicing and counting of cylindrical gels, and it facilitates the comparison of radioactive components in different samples on the same gel and of radioactive and unlabeled components. After the slab is fixed and excess labeled leucine is removed in a large volume of 10% TCA for 24 to 48 hours, it is stained if desired, and dehydrated by two 30-minute washes in four gel volumes of $Me_2SO$. It is then impregnated with fluor by soaking it 30 minutes in four gel volumes of $Me_2SO$ containining 20% (w/v) 2, 5-diphenyloxazole (PPO). The remaining PPO solution is carefully removed and added to one volume of distilled water, which precipitates the PPO; the PPO is recovered by filtration on a Buchner funnel. The gel is then washed overnight in distilled water. This rehydrates the gel, allowing it to be more easily dried, and precipitates the PPO within it. The PPO fluorescence stimulated by the weak β-particles of the tritiated proteins can now be detected on a piece of X-ray film. To do this, the gel is dried as described above, and a piece of X-ray film is taped to the edges of the gels, away from the sample channels. The gel and film are pressed between two glass plates wrapped in foil or black paper. The film is exposed by placing this package at $-70^{\circ}$ C, either in a low-temperature freezer or buried in dry ice. The X-ray film is then developed, and the relative amounts of radioactivity determined by scanning the film with a densitometer or a spectrophotometer.

## VIII. ERYTHROID DEVELOPMENT AND HEMOGLOBIN SYNTHESIS IN CELL CULTURE

Our culture system was adapted from that of Hagopian et al. [8]. A circle of vitelline membrane from an 18 to 36 hour embryo, including the blastoderm and embryo, is excised and placed in a Petri dish containing Howard's Ringer solution, which consists of 0.123M NaCl, 1.56 mM $CaCl_2$, and 5 mM KCl. The blastoderm and embryo are gently teased away from the vitelline membrane and removed to another Petri dish containing growth medium, which consists of Dulbecco's Modified Eagle's Medium, supplemented with antibiotics, fresh 2 mM glutamine, 10% dialyzed fetal-calf serum (GIBCO), 1% rooster serum (from a one-year-old white Leghorn rooster), 0.8 μg/ml of $FeSO_4$, 0.9 μg/ml of $ZnCl_2$, and 2 ng/ml $CuSO_4$. The Petri dish containing the blastoderms is swirled gently(at about 30 rpm) for ten minutes at room temperature to remove excess yolk. The yolky liquid is then carefully removed and replaced with 2 ml of fresh growth medium per embryo. The blastoderms are then dissociated into small fragments by sucking them in and out of a Pasteur pipette five to ten times, until visible fragements are dispersed. The suspension is divided among 35-mm tissue culture dishes (Falcon), each dish containing 1.5 ml of the suspension. The dishes are placed in a humidified incubator at 37$^{o}$ C in 5% carbon dioxide, 95% air.

We follow the development of erythroid cells in these cultures by mixing 50 to 100 μl aliquots of the suspended cells with 0.1 ml of 0.5% bovine serum albumin in NKM and centrifuging the cells against a microscope slide with a Shandon cytocentrifuge. These slides are fixed for five minutes in methanol, stained for twenty minutes in 2% Wright's stain--6% Giemsa stain in water, washed with water, and dried in air. Alternatively, the growth of cells, both attached and unattached, may be followed with phase optics with an inverted microscope. As the cells produce more hemoglobin, the refractive index of their cytoplasms increases and they darken.

During the first twenty-four hours of culture there are few unattached cells of any kind. By forty-eight hours of incubation, however, there are unattached cohesive groups of immature erythroid cells, some of which contain hemoglobin. As many as 4% of the erythroid cells are in mitosis. At seventy-two hours of incubation, up to 90% of the unattached cells contain hemoglobin, and very few of these cells are attached to the surface of the culture dish or to each other. After three days of culture the number of surviving cells decreases. In most of our experiments we begin with twenty-four hour blastoderms, usually the definitive streak or head process stage. When older embryos (up to four somites) are used as a source of cells, the progression of the culture is similar, with development corresponding to the total incubation time, i.e., the age of the starting embryo plus the time of incubation in culture.

We had originally assumed that the masses of cells seen in slides prepared with the cytocentrifuge were either artifacts of preparation or the result of the in vitro aggregation of erythroid precursors. We have been able, however, to demonstrate that the masses of cells observable after forty-eight hours of culture originate from smaller, tightly packed blastoderm fragments present and recognizable at the time of explanation. We have shown this by growing erythroid cultures in a warm air stream on the stage of an inverted phase-contrast microscope and photographing their development during a three-day period. To do this, we prepare 1.5 ml of a cell suspension in a 35-mm tissue culture dish, as described above, and equilibrate it with 5% carbon dioxide at 37° C for thirty minutes. We then seal the dish with Vaseline and place it on the stage of the inverted microscope. It is maintained at $37^{o}$ C by an air-stream incubator (Nicholson Precision Instruments). The culture dish was prepared beforehand by scoring a fine grid on its inner surface with a sterile wire screen. This allowed selected areas of the surface to be located and photographed repeatedly during the course of development.

Groups of erythroid cells derived from single fragments of blastoderm may be isolated after thirty-six to forty-eight hours of culture and subcultured to study their synthesis of individual hemoglobins. We do this by moving groups of about 50 immature erythroid cells with a micropipette, which has been drawn out to a tip diameter of 40 to 75 μm. Each such colony is placed in a 10-μl well of a Falcon multitest plate containing conditioned labeling medium and 10 μCi of [$^{3}$H]leucine. The labeling medium consists of 20% Dulbecco's Modified Eagle Medium, 80% leucine-free Minimal Essential Medium, supplemented with antibiotics, 2 mM fresh glutamine, 10% dialyzed fetal-calf serum (GIBCO), 1% rooster serum, 0.8 μg/ml of $FeSO_4$, 0.9 μg/ml of $ZnCl_2$, 2 ng/ml of $CuSO_4$, and nonessential amino acids (GIBCO). This medium is conditioned by growing in it an erythroid culture derived from twenty-four hour blastoderms at a concentration of one blastoderm per milliliter of labeling medium; after fourty-eight hours of culture all cells are removed by centrifugation.

The microcultures are allowed to grow for an additional twenty-four hours at 37° C in 5% carbon dioxide. Care must be taken not to allow the cultures to dry out. We do this by placing the multitest plate in a large Petri dish, the bottom of which is lined with wet paper. After the labeling is completed, the well contents are examined with the inverted microscope, and wells containing healthy-looking cells are collected and mixed with 10 μl of a suspension of about 2 x $10^8$ eight-day embryonic erythroid cells per milliliter of cold NKM containing 1 mM unlabeled leucine. These cells act as carrier cells during the washing of the cells and provide marker hemoglobins during analysis. Cells from an eight-day embryo contain both the early and the late sets of hemoglobins. The mixture is washed twice in cold NKM and lysed in 50 μl of TKM. Samples are then prepared and analyzed as described above.

## IX. ANALYSIS OF RNA SYNTHESIS IN ERYTHROBLASTS

In analyzing erythroblast's RNA metabolism, we employ the same techniques described by Jacobson (Chap. 3) and by Efstratiadis and Kafatos (Chap. 1). Particular problems arise, however, that derive from three properties of these highly differentiated, generally postmitotic cells: (a) their frequent contamination with leukocytes containing high levels of nucleases; (b) their relatively low level of RNA synthesis; and (c) the high concentration of hemoglobin in their cytoplasm.

To minimize the contamination of erythroblast lysates with leukocyte nucleases, we take two precautions: the careful removal of leukocytes (the "buffy coat" above the erythrocyte pellet) when washing the cells; and the rapid isolation of cytoplasm. When polysomes are not immediately extracted with phenol, they are brought to 0.5% SDS and frozen and stored in liquid nitrogen.

The low level of RNA synthesis in erythroblasts necessitates the use of relatively long incubation times. To study the rates of synthesis of nuclear and cytoplasmic RNAs, we incubate erythroblasts from anemic adults or from embryos of various ages in a medium containing $[^{14}C]$or $[^{3}H]$uridine for up to six hours. Cells are prepared, preincubated, and labeled as described for the analysis of in vitro hemoglobin synthesis, except that the growth medium is made 100 mM in adenosine, guanosine, and cytosine.

We use 1 μCi of [$^{14}$C]uridine (25 to 50 mCi/mmole) or 100 μCi of [$^{3}$H]uridine (20-40 Ci/mmole) per milliliter of cell suspension. Under these incubation conditions, synthesis of nuclear RNA continues unabated for at least five hours.

Total incorporation of labeled uridine into acid-insoluble material is monitored as a function of time by spotting aliquots onto filter paper as described above. The discs are processed in a batch by being placed successively into 100 to 200 ml of the following:

1. 10% TCA in 0.01M Na pyrophosphate (0-4° C), ten minutes
2. 5% TCA in 0.01M Na pyrophosphate (0-4° C), five minutes

   They are rinsed successively in 100 ml of the following:

3. 5% HAc, 2% HCl in acetone (0-4° C)
4. 95% EtOH
5. 100% EtOH
6. Three parts 100% EtOH, one part ether
7. Ether

They are then dried in air and counted in a toluene-based scintillation cocktail.

Incorporation of label into cytoplasmic RNA is determined after washing and lysing the cells at different times of incubation and removing the nuclei by centrifugation. Samples of the cell suspension are taken at appropriate times, and the cells are spun out and washed with ice-cold medium containing

1 mM unlabeled uridine. The cells are then lysed in 1 ml, or at least three volumes, of 1% Triton X-100 in TKM. After mixing for three minutes in ice, the nuclei are removed from the cytoplasm by centrifugation for fifteen minutes at 27,000 g and $4^{\circ}$ C. A sample of the cytoplasm is now added to at least five volumes of 10% TCA in 0.01M Na pyrophosphate. After ten minutes on ice, the precipitate is collected on a glass-fiber filter, washed with 5% TCA in 0.01M sodium pyrophosphate and acetone, dried, and counted in a toluene-based scintillation cocktail.

The large amount of hemoglobin in the cytoplasm of erythroblasts makes the isolation of protein-free RNA somewhat of a challenge. In preparing or analyzing mRNA, we therefore generally first isolate polysomes by differential centrifugation. To do this we clear the cytoplasm a second time by centrifugation for fifteen minutes at 27,000 g and $4^{\circ}$ C. The supernatant is then carefully removed, and polysomes are spun through a cushion of 40% sucrose in TKM at $4^{\circ}$ C either in a Spinco 40 rotor for 3.5 hours at 40,000 rpm through a 3 ml cushion or in a Spinco 60Ti rotor for 1.5 hours at 60,000 rpm through a 5 ml cushion. The polysomal pellet is washed in place with cold TKM and then either stored in liquid nitrogen or immediately dissolved in 1 ml of buffer containing 0.1M NaCl, 1 mM EDTA, 0.5% SDS, and 0.01M Tris-HCl (pH 7.4) (SDS buffer).

When we wish to analyze total cytoplasmic RNA, we first remove hemoglobin and other proteins by extracting the cleared cytoplasm with phenol-isoamyl alcohol-chloroform, as described by Perry et al. [30]. The sample to be extracted is brought to 0.5% SDS and 0.1M NaCl and extracted with two volumes of 50% phenol, 49% chloroform, and 1% isoamyl alcohol. Each extraction is performed by vigorously mixing the aqueous and organic phases for three minutes and separating the phases by centrifugation for three minutes at 2000 g. The organic (lower) phase is removed with a long Pasteur pipette, and the aqueous phase plus interphase is reextracted twice with two volumes of 2% isoamyl alcohol in chloroform. If a large interphase persists, the aqueous phase is vigorously mixed with one volume of liquified phenol. One volume of 2% isoamyl alcohol in chloroform is then added and mixed, and the phases are separated. The organic phase is removed, and the aqueous phase is reextracted with 2% isoamyl alcohol in chloroform until the interphase disappears, generally twice. The RNA is then precipitated from the aqueous phase with two volumes of ethanol. After at least two hours at $-20^{o}$ C, the precipitate is recovered by centrifugation at 27,000 g for fifteen minutes at -10° C, and dissolved in 1 ml of SDS buffer.

To correct for losses and to check for RNA degradation during phenol extraction, we generally add an external standard of about 20,000 cpm of *E. coli* [$^{14}$C]RNA to each sample to be extracted. This standard is prepared by growing a uracil-requiring strain of *E. coli* (B148) in the presence of [$^{14}$C]uracil. Bacteria are grown overnight at 37° C in 4x tryptone broth (DIFCO). A 0.5 ml aliquot of this cell suspension is added to 10 ml of VBC medium (see below), and the cells are recovered by centrifugation for ten minutes at 12,000 g and 4° C. The cells are then washed again in 10 ml of VBC medium and resuspended in about 10 ml of medium to give an $A_{550}$ of about 0.1. After one to two hours at 37° C, the cell number has doubled, and 20 μCi of [$^{14}$C]uracil (25-50 Ci/mole) is added. After being allowed to grow overnight at 37° C, the cells are harvested by centrifugation and lysed by vigorous mixing in 2 ml of 0.5% SDS, 0.02M NaAc (pH 5.2). This lysate is vigorously mixed with 4 ml of buffer-saturated phenol and incubated for five minutes at 68° C with the lysis buffer. The phases are again separated by centrifugation, and the two aqueous phases are combined. The RNA is precipitated with two volumes of EtOH and one-tenth volume of 2M NaCl. After being allowed to stand overnight at $-20^{\circ}$ C, the RNA is collected by centrifugation for 15 minutes at 27,000 g and $-10^{\circ}$ C and is dissolved in SDS buffer.

The VBC medium is prepared from a 2XVBC stock, casamino acids (DIFCO), and glucose. The 2XVBC stock is prepared by dissolving in 300 ml of distilled water 2 g of citric acid monohydrate, 10 g of $K_2HPO_4$, and 3.5 g of $NaH_2PO_4$. Then 2 g of $MgSO_4 \cdot 7H_2O$ is added, the pH is adjusted to 7.5, and the volume is brought to 500 ml. To make 100 ml of VBC medium, we add 1 g of glucose and 1 g of casamino acids to 50 ml of the 2XVBC and bring to volume with distilled water.

Erythroblast RNA can then be further analyzed by oligo(dT)-cellulose chromatography, centrifugation in sucrose gradients, and electrophoresis by standard techniques.

## ACKNOWLEDGMENTS

Many of the methods described in this paper derive from ones developed in Professor Vernon Ingram's laboratory. We are indebted to Professor Ingram and to Dr. Gail Bruns for many helpful discussions. We also thank Ms. Helen Hagopian and Ms. Marion Nadel,

Drs. Lloyd Waxman, Miles Paul, and Guido Guidotti for advice, and Ms. Kay Dickersin and Messrs. Patrick Kottas and Walter Taylor for technical assistance.

This work was supported by U.S.P.H.S. Grant #15885 and by a grant from the Milton Fund. H.V.C. was supported by U.S.P.H.S. Training Grant #T01-GM 00036 (to Harvard University) and K.S.P. was supported by U. S. P. H. S. Training Grant #T01-HD00415 (to Harvard University) and by a U.S.P.H.S. postdoctoral fellowship.

## REFERENCES

1. R.E. Lockard, and J. B. Lingrel, Biochem. Biophys. Res. Commun., 37, 204 (1969).
2. D. Housman, R. Pemberton and R. Taber, Proc. Natl. Acad. Sci. U.S.A., 68, 2716 (1971).
3. H. Aviv, and P. Leder, Proc. Natl. Acad. Sci. U.S.A., 69, 1403 (1972).
4. F. H. Wilt, Adv. Morphogenesis, 6, 89 (1967).
5. V. M. Ingram, Nature, 235, 338 (1972).
6. G. W. Settle, Contrib. Embryol., 241, 223 (1954).
7. H. K. Hagopian and V. M. Ingram, J. Cell. Biol., 51, 440 (1971).
8. H. K. Hagopian, J. A. Lippke, and V. M. Ingram, J. Cell. Biol., 54, 98 (1972).

9. M. Terada, L. Cantor, S. Metafora, R. A. Rifkind, A. Bank, and P. A. Marks, Proc. Natl. Acad. Sci. U.S.A., 69, 3575(1972).

10. J. E. Barker, J. A. Last, S. L. Adams, A. W. Nienhuis and W. F. Anderson, Proc. Natl. Acad. Sci. U.S.A., 70, 1793 (1973).

11. J. W. Adamson and G. Stamatoyannopoulos, Science, 180, 310 (1973).

12. K. Scherrer, L. Marcaud, F. Zajdela, I. M. London, and F. Gros, Proc. Natl. Acad. Sci. U.S.A., 56, 1571 (1966).

13. G. Attardi, H. Parnas, M-I.H. Hwang, and B. Attardi, J. Mol. Biol., 20, 145 (1966).

14. P. R. Williamson, and A. J. Tobin, manuscript in preparation.

15. E. S. Russell, in Regulation of Hematopoiesis I, A. S. Gordon, ed. Academic Press, New York, 1970, p. 649.

16. E. A. McCulloch, in Regulation of Hematopoiesis I, A. S. Gordon, ed. Academic Press, New York, 1970, p. 133.

17. T. H. J. Huisman, J. R. Adams, M. O. Dimmock, M. O. Edwards, E. E. and J. B. Wilson, J. Biol. Chem., 242, 2534 (1967).

18. T. G. Gabuzda, M. A. Schuman, K. K. Silver, and H. B. Lewis, J. Clin. Invest., 47, 1895 (1968).

19. G. A. P. Bruns, and V. M. Ingram, Phils. Trans. Soc. Lond. [Biol. Sci.], 266, 225 (1973).

20. A. M. Lucas, and C. Jamroz, Atlas of Avian Hematology, U.S.D.A. Monograph 25, Washington, D.C., 1961.

21. G. Brecher, Amer. J. Clin. Pathol., 19, 895 (1949)

22. D. L. Drabkin, and J. H. Asstin, J. Biol. Chem., 98, 719(1932).

23. O. Vesterberg in Methods in Enzymology, W. B. Jakoby, ed., Vol. XXII, Academic Press, New York, 1971, p. 389.

24. K. S. Pine, H. V. Colot, and A. H. Tobin, manuscript in preparation.

25. J. V. Maizel, in Methods in Virology, K. Maramorosch and H. Koprowski eds., Vol. V, Academic Press, New York, 1971.

26. A. Rossi Fanelli, E. Antonini, and A. Caputo, Biochim. Biophys. Acta, 30, 608 (1958).

27. R. T. Jones, Cold Spring Harbor Symp. Quant. Biol., 29, 297 (1964).

28. J. G. Schoenmakers and H. Bloemendal, Nature, 220, 790 (1968).

29. W. M. Bonner, and R. A. Laskey, Eur. J. Biochem., 46, 83(1974).

30. R. P. Perry, J. LaTorre, D. E. Kelley, and J. R. Greenberg, Biochim. Biophys. Acta, 262, 220 (1972).

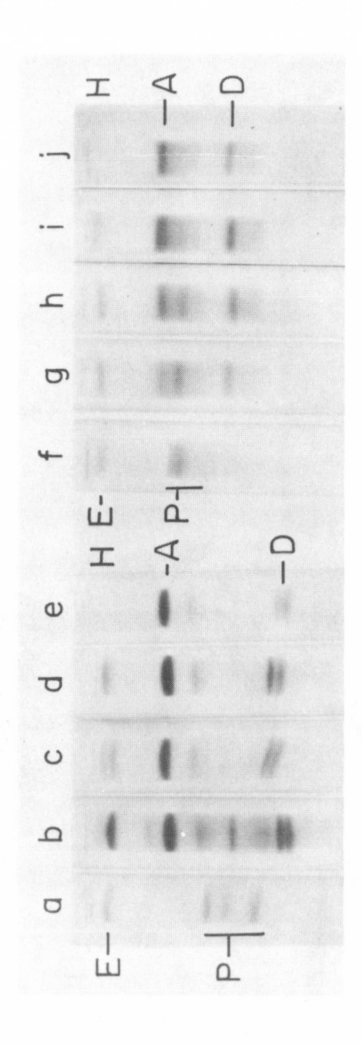


FIG. 1. Separation of chicken hemoglobins by isoelectric focusing and by gel electrophoresis.

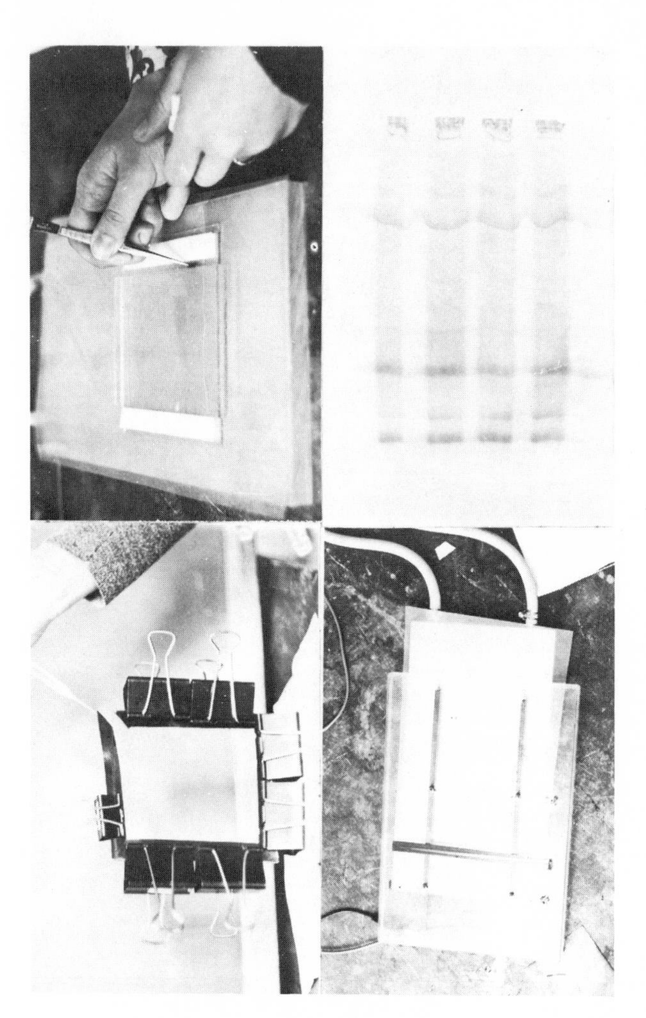

FIG. 2. Apparatus for acrylamide slab gel electrophoresis.

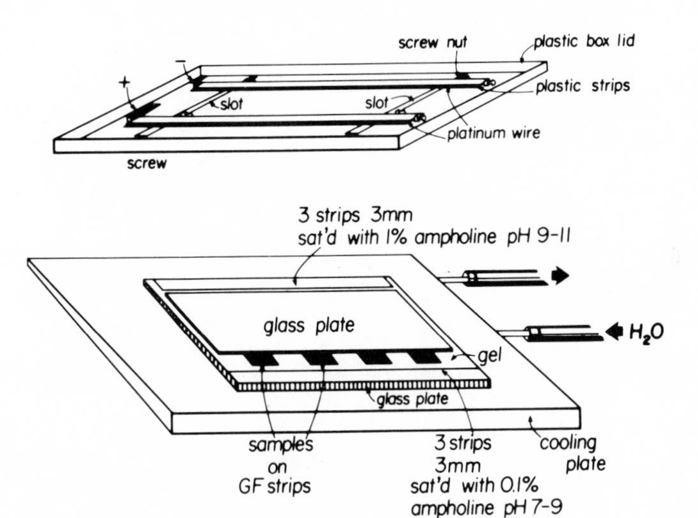


FIG. 3. Steps in the preparation of acrylamide slab gels for isoelectric focusing.

Chapter 5

# COLLAGEN AND LUNG GROWTH: A PROTOTYPE OF CONNECTIVE TISSUE DIFFERENTIATION

Morton J. Cowan, James F. Collins, and Ronald G. Crystal
Section on Pulmonary Biochemistry
National Heart and Lung Institute
Bethesda, Maryland

## I. INTRODUCTION

The mammalian lung is a heterogeneous organ composed of approximately 40 cell types and a complex connective tissue matrix [1,2]. Functionally, these components are separated into a vascular system carrying almost the entire cardiac output and an airway system composed of large branching conducting airways leading into small units called alveoli. In these units, the intimate relationship of the vascular and airway systems allows the exchange of oxygen and carbon dioxide between the blood and the atmosphere [3]. Although the lung has several "non-respiratory" functions [4,5], the major expression of the differentiated state of this organ is directed toward maximal efficiency in gas exchange.

The connective tissue matrix of lung is an essential part of this process. It maintains the structural integrity of the pulmonary vasculature, conducting airways and alveoli, and it has a major influence on the mechanical properties of the lung. Hence, it controls, to a large extent, the quantity and distribution of ventilation to the functioning gas exchange units [6]. Not only is the connective tissue matrix a major expression of lung differentiation, but there is compelling evidence that it strongly influences lung growth and eventual structure, therefore partially directing lung cellular differentiation [7-10]. Thus, the development and ultimate structure and function of lung is intimately related to the connective tissue comprising it.

In the adult lung, this matrix is composed (by weight) of collagen (60-70%), elastin (30-40%), and the amorphous ground substance, including proteoglycans ( 1%) [6,11-17]. Collagen, the major component of the lung connective tissue, may be used as a prototype of an experimental approach to investigating the role of connective tissue in defining the differentiated state of a vital organ. The areas which will be covered include: (a) methods to study lung collagen; (b) lung collagen heterogeneity; (c) changes in lung collagen content and synthesis in normal neonatal lung growth; and (d) the use of a model, in which the adult lung is induced to grow, to gain insight into the mechanisms controlling lung growth and differentiation.

## II. GENERAL ASPECTS OF COLLAGEN STRUCTURE, SYNTHESIS, AND ANALYSIS

### A. Collagen Biosynthesis

In order to appreciate methods used to study lung collagen it is necessary to understand how collagen is synthesized and subsequently processed to be included in the insoluble collagen fibril. Several recent reviews have described this process in detail [18-21]. In general, it appears that data on collagen biosynthesis and structure in other tissues is applicable to collagen in lung [12,22-24].

Although virtually nothing is known about the transcription of collagen mRNA, considerable effort has been made to define the translational and post-translational processes involved in collagen synthesis [18,20, 21, 25, 26]. It is currently believed that collagen is synthesized by polysomes on the endoplasmic reticulum, partially processed while still associated with the endoplasmic reticulum, and secreted into the extracellular space where further processing into the collagen fibril occurs. The product of collagen mRNA translation is a polypeptide of approximately 140,000 molecular weight, termed procollagen. The N-terminal portion of procollagen (the "pro" portion, 40,000 MW) is remarkable only for the presence of cysteine residues [18], whereas the C-terminal portion of procollagen (the collagen α-chain portion, 100,000 MW) has an unusual repeating sequence in which every third amino acid is glycine [19]. In addition to its glycine content, the α chain contains relatively large amounts of the imino acids, proline and hydroxyproline, and of alanine, aspartate, and glutamate. These six amino acids account for 80% of the total number. The chain also has small amounts of hydroxylysine, but no tryptophan [19]. Hydroxyproline and hydroxylysine are the result of the conversion of prolyl and lysyl residues after they are incorporated into the procollagen chain [18,20]. Since the "pro" portion of procollagen does not contain hydroxyproline or hydroxylysine, at least 20% of the polypeptide has to be synthesized before these residues appear. The conversion of proline to hydroxyproline and lysine to hydroxylysine requires the enzymes prolyl hydroxylase and lysyl hydroxylase, respectively. Both enzymes require atmospheric oxygen, ferrous iron, α-ketoglutarate, and a reducing agent such as ascorbic acid [18,20].

The pro portion of the pro-α chain is probably involved in the assembly of the individual chains into a unique, triple-stranded, helical structure termed protoprocollagen. This assembly is presumably promoted by disulfide links through cysteine residues in the pro portion [18, 27, 28]. As the triple-helical protoprocollagen molecule is processed, galactose, or glucose-galactose, is added to some hydroxylysine residues [29] and the protoprocollagen is secreted into the extracellular space. The "pro" portions are then cleaved in at least two stages by procollagen peptidases located extracellularly [18, 30]. The remaining helical molecule, called tropocollagen, is composed of three α chains and has a molecular weight of 300,000 [19].

The covalent incorporation of tropocollagen into the fibril depends on another extracellular enzyme, lysyl oxidase [31, 32], which catalyzes the deamination and aldehyde formation of lysine and hydroxylysine ε-amino groups. These aldehydes spontaneously form intra-and intermolecular cross-links [33, 34]. β-Aminopropionitrile (the lathyrogenic agent of sweet peas) inhibits lysyl oxidase activity and hence prevents covalent cross-link formation; it is frequently used experimentally to enhance the subsequent extraction of tropocollagen in neutral salt buffers and weak acid solutions [35, 36].

B. Identification of Collagen

A variety of techniques have been developed to identify collagen (Table 1). All are based on the unique physical and biochemical characteristics of this molecule. As will be discussed later, in relation to the qualitative and quantitative determination of lung collagen, each method has advantages and disadvantages in terms of convenience, specificity, and the state of the collagen after identification.

Early methods for the identification of collagen depended upon its insolubility in dilute base and the fact that, at neutral pH, collagen is readily converted to soluble gelatin at a well-defined temperature and under certain conditions [37]. This method is relatively crude and is not generally used.

Identification of the presence of hydroxyproline and/or hydroxylysine residues with an amino acid analyzer is a useful monitor for the presence of collagen [19]. Only two other proteins, elastin [19] and the Clq component of complement [38], contain hydroxyproline, and only Clq and collagen contain hydroxylysine [19,38]. Hydroxyproline is more commonly used because convenient, rapid colorimetric assays, which obviate the need for an amino acid analyzer, have been developed [39,40]. Since 9 - 12% of the residues of the α chain are hydroxyproline, whereas less than 1% of the amino acids in elastin and less than 4% of the amino acids in Clq complement are hydroxyproline, the presence of hydroxyproline is relatively specific for the presence of collagen [20,22]. No simple assays have been developed to detect hydroxylysine.

Collagen is the only protein known to be soluble in trichloroacetic acid at 90° C. This fact has been used in several assays to detect the presence of collagen [41]. Another useful method to detect the presence of collagen utilizes a clostridial collagenase, which degrades collagen at neutral pH in the presence of $Ca^{2+}$ [42]. This enzyme is readily available in a purified form and is specific for collagen.

There are several chromatographic procedures based on size (molecular sieve columns [43]) or charge (CM-cellulose [44] or DEAE-cellulose [44] ion-exchange resins) that can be used to identify and purify collagen in various forms (Table 1). Two electrophoretic methods, SDS-polyacrylamide gels [45] and acid polyacrylamide gels [46], are also widely used.

TABLE 1

Identification of Collagen[a]

| Method | Mechanism | Specificity for Collagen | State of Collagen: Prior to Assay | State of Collagen: After Assay |
|---|---|---|---|---|
| Gelatin formation [37] | Depends on pH, ions & temperature | Relatively crude | Intact, any form | Gelatin |
| Hydroxyproline [39,56] | Amino acid analyzer or pyrrole formation | Good; also present in elastin and Clq complement | Intact, any form | Hydrolyzed |
| Hydroxylysine [19] | Amino acid analyzer | Excellent, but present in small quantities | Intact, any form | Hydrolyzed |
| Trichloroacetic acid (90° C) [41] | Only collagen hydrolyzed | Good, but other small, noncollagen proteins may also be solubilized | Intact, any form | Hydrolyzed |
| Clostridial collagenase [42] | Enzymatic | Excellent | Intact, any form | Hydrolyzed |
| Molecular Sieve Chromatography [43] | Molecular weight | Combine with other methods unless used on previously purified material | Protoprocollagen, tropocollagen, β forms proα, α[b] | Same |

| | | | | |
|---|---|---|---|---|
| CM - cellulose [44, 93] | Charge | Excellent; collagen must be denatured | $\alpha$, p$\alpha$ ,$\beta$ | Same |
| DEAE - cellulose [44, 94] | Charge | Excellent; must be in triple-helical form | protoprocollagen, tropocollagen | Same |
| SDS-acrylamide gel electrophoresis [46] | Mostly size | Excellent, but collagen must be previously purified | $\alpha$,p$\alpha$, $\beta$ | SDS attached |
| Acidic acrylamide gel electrophoresis [46] | Size and charge | As above | As above | Difficult to extract from gel |
| CNBr peptide maps [19] | Size, charge | The best identification method; collagen must be previously purified | Intact, any form | Peptides |

[a] Although all of the identification methods may be used to identify unlabeled or labeled collagen, gelatin formation is usually used for unlabeled collagen, and the collagenase and TCA methods are usually used for labeled collagen.
[b] $\beta$ forms are two $\alpha$ chains covalently linked, only found in extracted collagen; p$\alpha$ , an intermediate in pro$\alpha$ to $\alpha$ conversion, can also be identified by this method.

As with other proteins, the absolute identification of collagen relies on peptide mapping procedures that separate portions of the α chains on the basis of size and/or charge in a characteristic and specific manner. Because collagen chains have few residues of methionine (usually less than ten), cyanogen bromide cleavage is most commonly used. The resulting CNBr peptides are then separated by size on molecular sieve columns or SDS-acrylamide gels or by charge on CM-cellulose or phosphocellulose columns [23,47, 48]. The use of CNBr mapping techniques has led to the discovery that mammalian collagen is not homogenous, but rather exists in several forms.

### C. Collagen Heterogeneity

Four types of tropocollagen have been described in mammalian tissues [19,49-52]. Each is composed of three α chains and each is synthesized in the protoprocollagen form [6,18]. Type I collagen has chain composition $[\alpha 1(I)]_2 \alpha 2$; Type II collagen chain composition $[\alpha 1(II)]_3$; Type III collagen chain composition $[\alpha 1(III)]_3$; and Type IV collagen (the so-called basement membrane collagen) chain composition $[\alpha 1(IV)]_3$. Each chain has a distinct and characteristic amino acid sequence as demonstrated by CNBr peptide mapping. As will be described below, the collagen of lung is the most heterogeneous in all organs studied, with all types being present.

The concept of collagen heterogeneity is of particular importance in a discussion of the role of connective tissue in organ differentiation. Although there were early suggestions that Type III collagen was a fetal form [53], more recent work has demonstrated Type III synthesis by adult tissues [49, 54]. Because there are specific structural differences in each collagen type, it is likely that the presence of each tropocollagen in an organ brings a specialized function to that tissue.

## III. METHODS TO STUDY LUNG COLLAGEN

### A. Collagen Already Present in Lung

In the ideal situation, it would be advantageous to take a specimen of lung and identify the type, quantity and location of all collagen present. This complete description depends on our ability to extract all collagen from lung. Unfortunately, a characterization of this type is not possible with current technology, primarily because of the physical properties and covalent crosslinks of collagen [6]. For example, although 15-20% of the total dry weight of adult lung is collagen, less than 1% of this collagen can be extracted in an intact form using neutral salt buffer, dilute acetic acid, or 5M guanidine (Table 2). The addition of β-aminopropionitrile to the animal's diet only increases this yield to approximately 5% (Table 2). Hence, it has not been possible to extensively identify the collagen already present in the lung. (For convenience, the collagen already present in lung will subsequently be called "unlabeled collagen." The newly synthesized collagen, detected by radioisotope methods, will be called "labeled collagen.") It is possible, however, to quantitate the total amount of unlabeled collagen in a lung or the concentration of unlabeled collagen per unit lung weight using hydroxyproline as a measure of collagen (Table 2). Provided that the hydroxyproline content of collagen is constant, it is estimated that 1 mg hydroxyproline is equivalent to 7.5 mg of lung collagen [12,22]. Although elastin also contains hydroxyproline, the relative amount of hydroxyproline in collagen compared with elastin and of collagen compared to elastin decreases the contribution of elastin to less than 1-3% of the total lung hydroxyproline.

The most convenient method for quantitation of unlabeled hydroxyproline involves the oxidation of hydroxyproline to a pyrrole [55, 56]. Problems have arisen with this method, due to incomplete oxidation, to incomplete removal of the oxidizing agent, and to the effect of competing amino acids [39]. These drawbacks have, to a large extent, been eliminated by the method of Prockop and Udenfriend [55], in which the pyrrole of hydroxyproline is selectively extracted with toluene, and oxidation is standardized by a measured excess of alanine.

In our experience, the use of the 90° C TCA-solubilization method or the clostridial collagenase method (Table 2) have not been useful in quantitating unlabeled lung collagen.

It is important to point out that these methods do not distinguish between types of collagen, and degrade collagen. Presently, quantitation of different types of collagen is not possible.

### B. Newly Synthesized Collagen

Most of the information concerning lung collagen has come from the study of newly synthesized, labeled collagen. Labeled collagen may be examined in short-term explant cultures [12,22,23], cell

TABLE 2
Methods to Study Lung Collagen
Unlabeled Collagen Already Present in Whole Lung or Lung Structure

| Method | Conditions | Yield | Use | Problems |
|---|---|---|---|---|
| Extraction with neutral salt, acetic acid, or guanidine [12] | Normal animal | 5% in fetus<br><1% in adult | Can extract enough collagen from fetus for partial characterization | Not quantitative; yield poor |
| Same as above [22] | Lathyritic animal fed βAPN | Approximately 5% in adult | Partial characterization | Same as above |
| Hydroxyproline [22] | Normal animal | 100% | Quantitation of total collagen or collagen/dry weight | Very little contamination by elastin; dependent on degree of hydroxylation. |
| TCA [22] | Normal animal | 100% | Same as above | Significant contamination with non-collagen proteins; difficult to quantitate |
| Collagenase [22] | Normal animal | Variable | Not useful--variability | |

Analysis of Newly Synthesized, Labeled Lung Collagen

| | | | | |
|---|---|---|---|---|
| [$^{14}C$] or [$^{3}H$] hydroxyproline [12,22, 24, 61] | Explant and culture | 100% | Quantitation of rate of collagen synthesis/ cell | See hydroxyproline [3] |
| [25] | Cell-free | Variable | For identification only | Must add enzyme and and cofactors to system after synthesis complete |
| Prolyl hydroxylase activity [57] | Explant and culture | Variable | For identification only | Does not necessarily correlate with collagen synthesis |
| Collagenase [22] | Explant | Variable | For identification only | Large amounts of other labeled & unlabeled proteins make assay difficult |
| [24,25] | Culture and cell-free | 100% | For quantitation of rate of collagen synthesis per cell (culture) or per polysome (cell free) | Enzyme must be pure and demonstrated to be specific |
| TCA (90° C [22,61] | Explant, culture, & cell-free | Variable | For quantitation of rates of collagen synthesis | Procedure also hydrolyzes labeled prolyl-tRNA |
| Molecular Sieve Chromatography [25] | Explant, culture, & cell free | Variable | For identification only; separates pro α from α | Carrier helps yield |

| | | | | |
|---|---|---|---|---|
| CM-cellulose [12,22,23,24] | Explant, culture, & cell free | Variable | For identification only; separates α1 from α2 chains | Precursor forms overlap α chains |
| DEAE - cellulose [24] | Culture | Variable | Excellent for identifying and separating protoprocollagen from tropocollagen; separates Type I from Type III | Must be run under specific conditions |
| SDS - acrylamide and acidic gel electrophoresis [22,25] | Explant, culture, and cell free | Variable | For identification of different forms (pro-tropocollagen vs. α chains) but not types of collagen | Cannot extract collagen from gel |
| CNBr peptide mapping [23,24] | Explant, culture, and cell free | Variable | For definitive identification of collagen chains; often used in combination with extraction of labeled collagen in neutral salt or acetic acid | Time consuming; marker peptides needed; not quantitative |

culture [24], and cell-free systems [25]. These techniques may be used for identification and quantitation of the collagen synthesized using several methods as outlined in Table 2. In general, the hydroxyproline and collagenase assays are most useful for quantitation of the rates of lung collagen synthesis, whereas chromatography, electrophoresis, and CNBr peptide mapping methods are most useful for qualitative determination of the types of collagen synthesized. Since the collagen is labeled with radioactive amino acids (often of high specific activity), very small quantities (pmole range) are sufficient to examine the collagen by chromatography or electrophoresis techniques.

The assay for labeled hydroxyproline is identical to that outlined above for unlabeled hydroxyproline (in fact, they are usually done simultaneously). The precursor amino acid must be [$^3$H] or [$^{14}$C]proline. A short column of silicic acid is used as a final step to remove any labeled proline that may have been extracted into the toluene [22, 56]. This assay can be used to quantitate the rate of collagen synthesis as long as the degree of hydroxylation of proline per collagen chain remains constant.

Prolyl hydroxylase activity has been used to

quantitate lung collagen synthesis [57]. In tissue culture systems, however, the level of activity of this enzyme has been shown to diverge from the rate of collagen synthesis per cell and, hence, its use is questionable [58].

The clostridial collagenase assay has been useful in explant systems from other organs [59], but, in our experience, the lung explant has too many labeled and unlabeled noncollagen proteins present to allow this assay to be reproducible. In lung tissue culture and cell-free systems, however, we have found it to be a convenient, accurate, and very reproducible assay. It not only allows quantitation of the rate of collagen synthesis but, by difference, measures the rate of noncollagen protein synthesis [25]. It is necessary, however, to have a purified, specific collagenase enzyme [42].

Although the usefulness of the 90° C TCA assay has been suggested by several studies [41,60], we find that it is difficult to use in lung systems. The major problems are that other noncollagen proteins become soluble in trichloroacetic acid under these conditions and that it is difficult to take into account the labeled prolyl-tRNA that is also hydrolyzed [25].

Molecular sieve chromatography provides a means for the purification of labeled collagen chains, as does CM-cellulose chromatography [12,22,23]. The SDS-acrylamide and acidic acrylamide gels have been useful only for identification of labeled collagen chains, particularly in separating triple-helical collagen from single chains, and in separating pro$\alpha$, p$\alpha$, and $\alpha$ chains [22, 24]. Interestingly, collagen $\alpha 2$ chains can be separated from $\alpha 1$ chains on SDS-acrylamide gels, even though their molecular weights are almost identical [19, 22].

DEAE-cellulose chromatography is extraordinarily useful in separating the Type III protoprocollagen, Type I protoprocollagen, and Type I tropocollagen found in the medium of lung cell cultures [24].

CNBr cleavage of purified $\alpha$ chains has been invaluable in demonstrating the heterogeneity of collagen chains synthesized by lung explants and lung cells in culture (see below).

Short-term Lung Explants

As little as 100 mg wet weight of lung tissue, minced with scissors into approximately 1 $mm^3$ pieces or sliced with a Stadie-Riggs slicer into 0.5-to-1.0 mm sections, can be used to quantitate rates of collagen synthesis or to qualitatively determine the kinds of collagen synthesized.

To quantitate the rates of collagen synthesis per cell, tissue slices or minces are incubated in equal volumes of Dulbecco's modified Eagle's medium and phosphate-buffered saline (PBS, pH 7.4) containing 0.5 mM ascorbic acid. 0.6 mM β-aminopropionitrile, and 0.012-0.024 mM [$^{14}$C] proline (Schwarz/Mann; 260 Ci/mole). The incubation is carried out at 37$^{o}$ C under a saturated atmosphere of 95% $O_2$ and 5% $CO_2$. At one-to-five hour intervals, samples are removed, washed with cold PBS, and homogenized in 0.5M HAC at 4$^{o}$ C. Aliquots are taken for determining dry weight, DNA, total protein, [$^{14}$C] and [$^{12}$C] hydroxyproline,&[$^{14}$C] proline in the tissue [22].

After a thirty minute lag period, presumably required for the labeled amino acid to enter the lung cells, the synthesis of collagen and noncollagen protein in these short-term explants is linear for more than five hours. The specific activity of [$^{14}$C]-proline in the tissue, used as an average approximation of the immediate precursor to protein synthesis, is constant throughout this period [22]. Extensive studies with this system have shown that the specific activity of the isotope in the tissue increases linearly with the concentration of isotope in the medium, and the rate of protein synthesis (including collagen) increases linearly with the specific activity of the isotope in the explant [22]. Hence, all rates of synthesis are expressed on the basis of the specific activity of [$^{14}C$]proline in the tissue [22].

The rate of collagen synthesis per cell is quantitated from [$^{14}C$]hydroxyproline per cell and expressed as nmoles [$^{14}C$]hydroxyproline/mg DNA·hr. The rate of noncollagen protein synthesis is quantitated from the [$^{14}C$]proline incorporated into noncollagen protein and expressed as nmoles [$^{14}C$]proline/mg DNA·hr. The percent collagen synthesis, representing the relative number of amino acids incorporated into all lung proteins per hour, can be easily calculated from these data [22]. It is necessary, however, to know the relative abundance of hydroxyproline in collagen compared to proline in noncollagen lung protein (2.06 for rabbit lung) [22,61].

The explant system can also be used to determine qualitatively the types of collagen synthesized by slices or minces of whole lung or specific lung structures, such as blood vessels or the tracheobronchial tree. Neutral salt or dilute acetic acid extracts approximately 50% of the newly synthesized (labeled) collagen from these explants. This newly synthesized material is then purified by CM-cellulose chromatography (Fig. 1A) and cleaved with CNBr prior to peptide mapping on SDS-acrylamide gels or CM-cellulose columns. Recent studies have demonstrated that 100% of the collagen synthesized in these explants can be extracted in SDS-dithiothreitol-Tris buffers or 8M urea-dithiothreitol-Tris buffers, but the presence of SDS or urea makes further analysis cumbersome.

Lung Cell Culture

It is possible to disperse fetal or neonatal lung tissue into individual viable cells that can be maintained in culture over several passages. Colony clone or single-cell clone techniques can be used to maintain a single cell type (e.g., WI-38, a diploid fibroblast from human fetal lung). An easier method is to allow all cells to grow without selecting a specific cell type [24]. Although the latter often take on the appearance of fibroblasts by phase microscopy, electron microscopic analysis has shown that several cell types are often present.

These systems can be used to quantitate rates of collagen synthesis and rates of noncollagen protein synthesis per cell, utilizing either the hydroxyproline or the collagenase methods (Table 2). From these data, the relative rate of collagen synthesis can be calculated (see above).

Lung cell culture has also proved useful in qualitatively determining the types of collagen synthesized by each cell type [24].

The great promise of such systems is the study of lung cell differentiation under varying conditions including cell-cell interactions with alveolar macrophages, lymphocytes, and other cell types.

## Lung Cell-Free Systems

A cell-free system that synthesizes lung protein including collagen is essential to appreciate the molecular mechanisms involved in the changes in collagen accumulation during growth and development. The in vitro synthesis of collagen in cell-free systems has usually been accomplished with polysomes [62-64] or mRNA [26,65,66] purified from embryonic chick bone, a tissue that makes a large amount of collagen and from which subcellular components may be readily prepared. The development of a lung cell-free system has been difficult, since only 4 to 5% of the protein synthesized by rabbit [22] and human [12] lung is collagen. In addition, the increase in the amount of lung connective tissue with age makes it difficult to prepare subcellular components. But, the promise of answering questions about control mechanisms of lung collagen biosynthesis has prompted the development of an active, heterologous, cell-free, protein-synthesizing system [25].

The heterologous, cell-free system for studying lung collagen synthesis consists of lung polysomes, rabbit liver tRNA, and 0.5M KCl ribosomal wash fraction from rabbit reticulocytes. The wash fraction contains all the factors required for initiation, elongation, and release of polypeptide chains, as well as all of the synthetases necessary for tRNA acylation. [67]. Free (cytoplasmic) and bound (derived from the endoplasmic reticulum) lung polysomes can be separated and purified by differential centrifugation (Fig. 2), and the other components are prepared by standard techniques [25].

There are several features that seem to be characteristic of lung polysome preparations. There is a predominance of single ribosomes in lung polysomes compared to profiles for polysomes from other tissues prepared under similar conditions. Whether this is the actual state of the polysomes in the tissue, or an artifact secondary to isolation procedures, is not known. In addition, bound polysomes from lung (Fig. 3A) have a greater proportion of large polysomes than free polysomes (Fig. 3B). This might be important in relation to lung collagen synthesis since collagen mRNA is very large and, hence, has the potential for accommodating more ribosomes [26].

Total protein synthesis in the heterologous cell-free system is determined by quantitating the incorporation of labeled proline into trichloroacetic acid-precipitable protein. Collagen synthesis is quantitated with the collagenase assay [25]. The usual method of destroying labeled aminoacyl-tRNA (exposure to trichloroacetic acid at 90° C) cannot be used in this system because such exposure hydrolyzes collagen as well as the prolyl-tRNA. This problem is circumvented by breaking the ester bond between the labeled amino acid and the tRNA by exposure to pH 10 for 30 minutes before trichloroacetic acid precipitation to determine total protein synthesis [25, 62].

Bound and free polysomes are comparably active in total protein synthesis and require ribosomal wash fraction, ATP, GTP, and tRNA for maximal synthesis (Table 3). Amino acid incorporation is partially inhibited by the inhibitors of polypeptide chain initiation, sodium fluoride, and aurintricarboxylic acid, and completely by puromycin. Results of experiments involving the donation of formyl-[$^{35}$S]methionine from formyl-[$^{35}$S]methionyl-$tRNA_f$, the mammalian initiator tRNA, indicate that the system is initiating, as well as elongating, polypeptide chains [25].

TABLE 3

Dependency of the Cell-Free System on Various Components

| | pmoles [$^{14}$C]proline incorporated per $A_{260}$ polysomes in 18 min. | |
|---|---|---|
| | Bound polysomes | Free polysomes |
| Complete system[a] | 20.9 | 19.5 |
| - polysomes | 1.3 | 0.6 |
| - wash fraction | 0.3 | 0 |
| - tRNA | 1.9 | 0.9 |
| - GTP | 14.8 | 13.4 |
| - ATP | 4.2 | 2.0 |
| + 1 mM puromycin | 1.3 | 0 |
| + 10 mM NaF | 9.4 | 6.9 |
| + 0.2 mM aurintricarboxylate | 11.7 | 8.5 |

[a]A 100-μl reaction mixture contained 20 mM Tris-HCl (pH 7.4), 4 mM dithiothreitol, 1 mM ATP, 0.2 mM GTP, 0.1 I.U. pyruvate kinase, 3 mM phosphoenolypyruvate, 5 mM $MgCl_2$, 100 mM KCl, 0.27 $A_{260}$ units of rabbit liver tRNA, 0.15 $A_{260}$ units of lung bound or free polysomes from 1-week-old rabbits, 160 μg of rabbit reticulocvte 0.5M KCl ribosomal wash fraction, 40 μM 19 amino acids (except proline), and 40 μM [$^{14}$C]proline. A reaction mixture that contained neither polysomes nor wash fraction was considered to have zero incorporation.

Whereas sensitivity to collagenase provides a simple and rapid measurement of lung collagen synthesized by the cell-free system, the collagen product has also been demonstrated by such methods as hydroxylation by peptidyl prolyl hydrocylase, CM-cellulose chromatography (Fig. 1B), and SDS-acrylamide gel electrophoresis [25]. As seen in Figure 1, B the collagen product of the cell-free system is qualitatively similar to the product made in the lung explant system (Figure 1A). Both systems have prominent α1 and α2 peaks. As noted by other investigators using other systems, the heterologous cell-free system produces very little collagenase-sensitive material that has a molecular weight over 100,000 [26, 65]. This is probably due to a combination of the difficulty in handling procollagen [68], the absence of procollagen in the collagen carrier used, the small amount of material made in the cell-free system, and the possible presence of proteases in the cell-free system.

The optimization of a cell-free system for a number of major components is important in order to maximize collagen-synthesizing activity and to assure that the system is dependent only upon lung polysome fraction. Several components can be varied until conditions are found in which both total protein synthesis and collagen synthesis are maximal and the percent of total protein synthesis which is represented by collagen synthesis (percent collagen synthesis) is relatively independent of minor changes in component concentration (Fig. 4). Amino acid incorporation and collagen synthesis are maximal at 5 mM $MgCl_2$(A), 90 to 140 mM KCl (B), above 100 μg of reticulocyte 0.5M KCl wash fraction (C), and at 40 μM amino acids (D). Total incorporation and collagen synthesis are linear for 20 to 25 minutes (E), and up to at least 0.3 units of polysomes/100 μl reaction mixture (F). It is of interest that total protein synthesis is maximal with 50 μg of reticulocyte ribosomal wash fraction (C), whereas collagen synthesis requires at least twice as much wash fraction to achieve maximal synthesis. This finding suggests that there may be differential requirements of collagen synthesis and total protein synthesis for initiation, elongation, and other factors in the ribosomal wash fraction.

## IV. HETEROGENEITY OF LUNG COLLAGEN

Most of the work on the heterogeneity of collagen chains in tissues other than lung has utilized unlabeled collagen. As discussed in the methods section, however, this is very difficult to accomplish in lung, although we have been able to extract small amounts of predominantly Type I collagen from fetal-human lung and from adult lathyritic-rabbit lung [12,22]. The most useful technique to investigate lung collagen heterogeneity has been to combine collagen synthesis methods (both explant and cell culture) with the more classical CNBr peptide mapping methods [23,24]. Thus, all of the information on lung collagen heterogeneity does not describe the collagen present in the tissue, but rather the types of collagen being synthesized by lung at that time. As discussed above, this information is qualitative and not quantitative.

Three of the four mammalian collagen types have now been described in lung with almost complete CNBr peptide maps (Table 4). Type I collagen is synthesized by peripheral lung [23], the vascular tree [23], and by lung fibroblasts [24]. Type II collagen is synthesized by trachea [23]and by the bronchial tree [23]. Type III collagen is synthesiaed by lung fibroblasts [24] and it may be synthesized by adult human lung [12]. Although Type IV collagen has not been well described in lung, the presence of a large alveolar basement membrane (70 $m^2$ in the adult human lung) and the presence of 3-hydroxyproline (considered to be a "marker" for Type

TABLE 4
Lung Collagen Heterogeneity

| Structure or Cell Type | Tropocollagen | Type | Analogous Source | Reference |
|---|---|---|---|---|
| Tracheobronchial tree | $[\alpha 1(II)]_3$ | II | Cartilage | 23,50 |
| Vascular tree | $[\alpha 1(I)]_2\alpha 2$ | I | Skin | 19,23 |
| | $[\alpha 1(III)]_3$ ? | III | Aorta | 49 |
| Peripheral lung | $[\alpha 1(I)]_2\alpha 2$ | I | Skin | 19,22,23,94 |
| | $[\alpha 1(III)]_3$ ? | III | Aorta | 49 |
| | $[\alpha 1(IV)]_3$ ? | IV | Renal basement membrane | 52 |
| Fibroblast | $[\alpha 1(I)]_2\alpha 2$ | I | Skin fibroblast | 24,95 |
| | $[\alpha 1(III)]_3$ | III | Skin fibroblast | 24,95 |

IV collagen) in extracts of rabbit [22] and human [12] lung suggest that this collagen type is present.

Considering the lung as a whole, the expression of its differentiated state in regards to collagen heterogeneity is indeed complex. It is not known at what stages in development each collagen type is synthesized nor what role each collagen type plays in determining the mechanical properties of the lung. It is likely that of the 40 cell types present, several will be implicated as being responsible for collagen synthesis. For example, in culture, human fetal lung fibroblasts synthesize both Type I and Type III collagen [24], and initial studies with cat lung epithelial cells have demonstrated collagen synthesis. It is most likely that in the tracheobronchial tree, the chondroblast [69] is responsible for Type II collagen synthesis. In the future it should be possible, using cell culture, protein synthesis, and immunologic techniques, to completely define the quantities and types of collagen synthesized by lung cells at different stages in lung growth.

## V. CHANGES IN LUNG COLLAGEN CONTENT AND SYNTHESIS IN NORMAL NEONATAL LUNG GROWTH

An inbred strain of New Zealand white rabbits was used in all of our studies of collagen and lung growth. The lungs of these animals are free from infection and they are of the appropriate size to provide sufficient material for explant and cell-free studies. Histologic analysis suggests that, as in the human lung, development of the rabbit lung also involves a stage of postnatal alveolar maturation [70-73].

A. Total Amount and Density of Lung Collagen

The total lung dry weight in a rabbit fetus (-10 days gestation) increases over 30-fold by 60 days of age. The density of lung collagen (relative to lung dry weight) increases 5-fold throughout the fetal and neonatal periods and is relatively constant after two months of age. This finding is consistent with the fact that the collagen content of human lung does not vary in adult life [14]. The neonatal changes in collagen content may, however, be one aspect of the maturation process of the distal airways [73-75].

The density of DNA (per tissue dry weight) in fetal rabbit lung is higher than in the neonatal and adult lung. This change in the perinatal period may reflect the emergent functional responsibility of the newborn lung with a stabilization of the relative extracellular mass. Since the number of lung cells per unit lung mass is constant for the newborn to adult growth period, but the total dry weight increases tenfold, the total number of cells in the postnatal lung also increases tenfold. The types and localization of cells responsible for this increase are only beginning to be described [76].

Studies of changes in collagen and DNA content and density in human fetal and adult lung are similar to those seen in rabbit lung, although they are more pronounced [12].

B. Quantitation of Collagen Synthesis During Growth

From late fetal to two months of age, the rate of collagen synthesis per cell decreases over eightfold. In comparison, the decrease in total protein synthesis per cell is approximately fivefold, but at a much more rapid rate. Because collagen synthesis changes less rapidly, the percentage of total protein synthesis that is represented by collagen synthesis peaks during the first week of life and then reverts to adult levels by three months of age. This marked increase in the relative emphasis on collagen synthesis occurs during a period when significant accumulation of collagen in the lung (both total amount and density) is occurring. Preliminary studies of lung collagen proteolysis have suggested that newly synthesized lung collagen is broken down at a relatively constant level throughout growth. As techniques are developed to quantitate the proteolysis of newly synthesized and unlabeled lung collagen, it should be possible to determine the relative contributions of synthesis and proteolytic processes in lung collagen accumulation. In the early neonatal period, there is a significant increase in lung cell division in mice, hamsters, and rats [76, 77]. In the alveoli, this increase in cell division is seen in endothelial and alveolar epithelial cells, but is most pronounced in septal fibroblasts [76]. Whether this increase in the relative number of connective tissue cells might explain the increase in percent collagen synthesis noted before birth will have to await further study.

### C. Control of Collagen Synthesis During Growth

If it is true, as suggested by the explant experiments, that the rate of lung collagen synthesis is an important factor in the accumulation of collagen in lung during growth, then it should be possible to use cell-free systems to delineate the mechanisms responsible for these changes.

A possible explanation for the rapid decrease in the rate of collagen synthesis, as well as the percent of collagen synthesis after birth, is that this decrease represents a change in lung cell differentiation, so that the total population of lung cells synthesize, on the average, less collagen relative to other proteins. This may be expressed on the transcriptional (gene "switching") level, translational level, or post-translational level. Experiments with the heterologous cell-free system described above have eliminated post-translational mechanisms as important contributing factors in the changes seen in percent collagen synthesis with growth (Fig. 5).

It is important to note that the cell-free data are presented as rate of synthesis per polysome rather than per cell as in the explant studies. The number of polysomes recovered per unit lung mass does not change with age [25], nor does the number of cells per unit mass, so the different techniques are comparable. Likewise, the relative number of bound polysomes to free polysomes recovered is relatively constnat (approximately 10 to 20%).

Although total lung protein synthesis per polysome is the same for free and bound polysomes (Fig. 5A), collagen synthesis per polysome is low on the free and high on the bound polysomes (Fig. 5B). Hence, whatever changes occur in lung collagen synthesis on the subcellular level should be reflected on the bound polysomes. As in the explants, collagen synthesis directed by bound polysomes decreases significantly during fetal to adult growth. The lack of marked change in total protein synthesis is reflected in the percent collagen synthesis (Fig. 5C), which is halved during this period.

Thus, the bound polysomes of lung appear to be directing the decrease in the relative and absolute rate of collagen synthesis seen during neonatal lung growth. Since the polysomes are the only component in the cell-free system derived from lung, this suggests that the mechanisms operant are not post-translational and probably not translational (unless a controlling "factor" is associated with lung polysomes). This concept is supported by the fact that the changes in percent collagen synthesis that occur between fetal and two-month-old rabbit lungs are similar if lung tRNA replaces liver tRNA or if lung ribosomal wash fraction replaces reticulocyte ribosomal wash fraction. It is still possible that there are translational mechanisms important in the control of collagen synthesis (e.g., initiation factors, tRNA), but these do not appear to be major determining factors in the changes noted. The development of a homologous system with all components from lung tissue will permit the determination of whether there are other factors present in the different age animals that can further increase the difference in collagen synthesis between the fetal and two-month animals.

## VI. A MODEL OF LUNG GROWTH IN THE ADULT

During normal lung maturation in rabbits, complex changes occur in collagen content and synthesis. But, by two months of age, collagen synthesis and accumulation are near adult values. Morphologically, the lung of the two-month-old rabbit is similar to the adult [70]. This pattern of lung growth can be altered when animals are exercised, exposed to hyperoxia or hypoxia, or given growth and thyroid hormones [78-83]. It has also been demonstrated in several species that when one lung is removed the remaining lung nearly doubles in volume, weight, and number of cells [61,84-87].

A. Lung Growth Induced by Unilateral Pneumonectomy

In a ten-week-old rabbit, if the left lung is removed, the remaining lung (Fig. 6B) is grossly larger than the control (Fig. 6A) by 28 days after surgery. This stimulated growth of the remaining lung is, in part, represented by significant increases in the total right lung dry weight (Fig. 7A), DNA (Fig. 7B) and collagen (Fig. 7C) by the 14th post-operative day, with a near doubling as compared to controls by the 28th day.

Although the remaining lung in the pneumonectomized animal shows an increase in total number of cells and amount of collagen, the density of cells and of collagen remains constant at control levels [61]. This finding is compatible with the light microscopic picture of lung sections from pneumonectomized animals showing no significant difference in connective tissue arrangement or cellular content from controls [61].

The accumulation of collagen in the lung after pneumonectomy is preceded by an increase in the rate of collagen synthesis relative to lung total protein synthesis (Fig. 7). By 14 days after surgery, when the total collagen accumulation in the lung is leveling off (Fig. 7C), the rate of lung collagen synthesis is returning to control values.

The twofold increase in the percent collagen synthesis in the right lung on the seventh day after left pneumonectomy is qualitatively similar to the threefold increase seen in the newborn rabbit lung. Total DNA, total collagen, and density of DNA in normal neonatal lung growth and lung growth after pneumonectomy also show similar patterns.

Although the overall pattern of pneumonectomy-induced lung growth in adults appears to be identical to normal neonatal lung growth, there is no change in collagen density in pneumonectomy-induced growth as compared to the marked increase in collagen density seen in neonatal growth. Hence, although there is histologic, physiologic, and biochemical evidence that lung growth after pneumonectomy in adults is remarkably similar to normal lung maturation, there are some biochemical data suggesting that the mechanisms controlling growth may be different.

The types of collagen synthesized during lung growth after pneumonectomy are not known, although preliminary evidence suggests that at least $\alpha 1(I)$ and $\alpha 2$ chains are involved [61]. As in normal lung growth, the specific cell types responsible for the changes in collagen synthesis and the specific collagens synthesized need to be studied.

### B. Control of Post-Pneumonectomy Lung Growth

When the chest cavity is opened in an animal that has undergone unilateral pneumonectomy, it is apparent that the enlarged lung occupies a significant portion of the entire chest cavity. If the empty left chest cavity (resulting from left pneumonectomy) is filled with wax ("plombage") immediately after pneumonectomy, the growth response of the right lung is significantly blunted (Fig. 7).

The repression of lung collagen accumulation and synthesis after pneumonectomy by wax plombage implies that some of these biochemical changes may be initiated by the mechanical stimulus of providing a larger space into which the lung may grow. This is not surprising, since complex mesenchymal-epithelial interactions are known to occur in normal fetal lung maturation, and mechanical disruption of these interactions results in abnormal lung development [8-10,88-92].

Although the intricate branching and shaping of the lung during embryogenesis is certainly due to genetic direction, the control may be, in part, exerted through limitations imposed by the chest wall (and hence chest cavity size), rather than endogenously by means of factors in the lung tissue.

Lung growth after pneumonectomy will have to be examined with respect to replicative, translational, and post-translational mechanisms to fully understand its relationship to normal lung growth. It is interesting to speculate, however, how mechanical factors may have significant influence on the expression of lung differentiation.

## VII. SUMMARY

It has been our purpose to describe a technology that is applicable to study a specific differentiated expression of a complex mammalian organ. Although it is apparent that the description and control of lung growth is complex, an understanding of lung collagen synthesis during maturation should yield significant insights into gene expression in mammalian systems. The technology described emphasizes the overall pattern of lung collagen content and synthesis. The next generation of studies will be aimed at reducing the variables involved by using model systems, particularly tissue culture, of lung growth and differentiation.

REFERENCES

1. C. Kuhn, III, in The Biochemical Basis of Lung Function, R. G. Crystal, ed., Marcel Dekker, New York, 1975.
2. S. P. Sorokin, in Morphology of Experimental Respiratory Carcinogenesis, P. Nettsheim, M. G. Hanna, Jr., and J. W. Deatherage, Jr., eds., U.S. Atomic Energy Commission, Oak Ridge, 1970, p.3.
3. C. Nagaishi, Functional Anatomy and Histology of the Lung, University Park Press, Baltimore, 1972.
4. D. F. Tierney, Ann. Rev. Physiol., 36, 209 (1974).
5. H. O. Heinemann and A. P. Fishman, Physiol. Rev., 49, 1 (1969).
6. A. J. Hance and R. G. Crystal, in The Biochemical Basis of Lung Function, R. G. Crystal, ed., Marcel Dekker, New York, 1975.
7. R. L. Trelstad, J. Histochem. and Cytochem., 21, 521 (1973).
8. B. S. Spooner and N. K. Wessells, J. Exp. Zool., 175, 445 (1970).
9. N. K. Wessells, J. Exp. Zool., 175, 455 (1970).
10. T. Alescio and E. C. Piperno, J. Embryol. Exp. Morph., 17, 213(1967).
11. R. G. Crystal, Fed. Proc., 33, 2248 (1974).

12. K. Bradley, S. McConnell-Breul, and R. G. Crystal, J. Clin. Invest., in press (1975).

13. A. L. Horwitz, N. A. Elson, and R. G. Crystal, in The Biochemical Basis of Lung Function, R. G. Crystal, ed., Marcel Dekker, New York, 1975.

14. J. A. Pierce and J. B. Hocott, J. Clin. Invest., 39, 8 (1960).

15. J. A. Pierce and R. V. Ebert, Thorax, 20, 469 (1965).

16. R. Johnson and F. A. Andrews, Chest, 57, 239 (1970).

17. R. Johnson and J. Thomas, Biochem. J., 127, 261 (1972).

18. P. Bornstein, Ann. Rev. Biochem., 43, 567 (1974).

19. P. M. Gallop, O. O.Blumenfeld, and S. Seifter, Ann. Rev. Biochem., 41, 617 (1972).

20. M. E. Grant and D. J. Prockop, N. Engl. J. Med., 286, 194, 242, 291 (1972).

21. R. E. Priest, in Molecular Pathology of Connective Tissues, R. Perez-Tamayo and M. Rojkind, eds., Marcel Dekker, New York, 1973, p. 105.

22. K. H. Bradley, S. D. McConnell, and R. G. Crystal, J. Biol. Chem., 249, 2674 (1974).

23. K. Bradley, S. McConnell-Breul, and R. G. Crystal, Proc. Natl. Acad. Sci. U. S. A., 71, 2828 (1974).

24. A. J. Hance, K. Bradley, and R. G. Crystal, submitted for publication.

25. J. F. Collins and R. G. Crystal, J. Biol. Chem., in press (1975).

26. H. Boedtker, R. B. Crkvenjakov, J. A. Last, and P. Doty, Proc. Natl. Acad. Sci. U. S. A., 71, 4208 (1974).

27. B. Goldberg, E. H. Epstein, Jr., and C. J. Sherr, Proc. Natl. Acad. Sci. U. S. A., 69, 3655 (1972).

28. C. J. Sherr and B. Goldberg, Science, 180, 1190 (1973).

29. A. M. Robert, B. Robert and L. Robert, in Chemistry and Molecular Biology of the Intercellular Matrix, E. A. Balazs, ed., Academic Press, London, 1970, p. 237.

30. C. M. Lapiere, A. Lenaers, and L. D. Kohn, Proc. Natl. Acad. Sci., U. S. A., 68, 3054 (1971).

31. R. C. Siegel, S. R. Pinnell, and G. R. Martin, Biochemistry, 9, 4486 (1970).

32. S. R. Pinnell and G. R. Martin, Proc. Natl. Acad. Sci. U. S. A., 61, 708 (1968).

33. M. L. Tanzer, Science, 180, 561 (1973).

34. A. J. Bailey, S. P. Robins, and G. Balian, Nature, 251, 105 (1974).

35. C. I. Levene, in Molecular Pathology of Connective Tissues, R. Perez-Tamayo and M. Rojkind, eds., Marcel Dekker, New York, 1973, p. 175.

36. E. J. Miller, G. R. Martin, K. A. Piez, and M. J. Powers, J. Biol. Chem., 242, 548 (1967).

37. O. H. Lowry, D. R. Gilligan, and E. M. Katersky, J. Biol. Chem., 139, 795 (1941).

38. K. Yonemasu, R. M. Stroud, W. Niedermeier, and W. T. Butler, Biochem. Biophys. Res. Commun., 43, 1388 (1971).

39. D. S. Jackson and E. G. Cleary, Methods Biochem. Anal., 15, 25 (1967).

40. R. E. Neuman and M. A. Logan, J. Biol. Chem., 184, 299 (1950).

41. R. O. Langner and R. A. Newman, Anal. Biochem., 48, 73 (1972).

42. B. Peterkofsky and R. Diegelmann, Biochem., 10, 988 (1971).

43. K. A. Piez, Anal. Biochem., 26, 305 (1968).

44. B. D. Smith, P. H. Byers, and G. R. Martin, Proc. Natl. Acad. Sci. U. S. A., 69, 3260 (1972).

45. K. Weber and M. Osborn, J. Biol. Chem., 244, 4406 (1969).

46. T. Sakai and J. Gross, Biochemistry, 6, 518 (1967).

47. E. J. Miller, J. M. Lane, and K. A. Piez, Biochemistry, 8, 30 (1969).

48. H. Furthmayr and R. Timpl, Anal. Biochem., 41, 510 (1971).

49. E. Chung and E. J. Miller, Science, 183, 1200 (1974).

50. E. J. Miller, Biochemistry, 10, 1652 (1971).

51. R. L. Trelstad, A. H. Kang, B. P. Toole, and J. Gross, J. Biol. Chem., 247, 6469 (1972).

52. N. A. Kefalides, Biochem. Biophys. Res. Commun., 45, 226 (1971).

53. E. J. Miller, E. J. Epstein, Jr., and K. A. Piez, Biochem. Biophys. Res. Commun., 42, 1024 (1971).

54. E. H. Epstein, Jr., J. Biol. Chem., 249, 3225 (1974).

55. D. J. Prockop and S. Udenfried, Anal. Biochem., 1, 228 (1960).

56. K. Juva and D. J. Prockop, Anal. Biochem., 15, 77 (1966).

57. J. Halme, Biochim. Biophys. Acta, 192, 90 (1969).

58. H. Green and B. Goldberg, Cultured Cells, Symposium of the International Society for Cell Biology, 7, 123 (1968).

59. R. F. Diegelmann and B. Peterkofsky, Dev. Biol., 28, 443 (1972).

60. B. Peterkofsky and S. Udenfried, J. Biol. Chem., 238, 3966 (1963).

61. M. J. Cowan and R. G. Crystal, Amer. Rev. Resp. Dis., in press (1975).

62. S. S. Kerwar, L. D. Kohn, C. M. Lapiere, and H. Weissbach, Proc. Natl. Acad. Sci. U. S. A., 69, 2727 (1972).

63. E. Lazarides and L. N. Lukens, Nature (New Biol.), 232, 37 (1971).

64. R. F. Diegelmann, L. Bernstein, and B. Peterkofsky, J. Biol. Chem., 248, 6514 (1973).

65. K. Benveniste, J. Wilczek, and R. Stern, Nature, 246, 303 (1973).

66. K. Benveniste, J. Wilczek, and R. Stern, Fed. Proc., 33, 1441 (1974).

67. R. G. Crystal, N. A. Elson, and W. F. Anderson, Methods Enzymol., 30, 101(1974).

68. G. Bellamy and P. Bornstein, Proc. Natl. Acad. Sci. U. S. A., 68, 1138 (1971).

69. R. Mayne, J. F. Schlitz, and H. Holtzer, in Biology of Fibroblast, E. Kulonen and J. Pikkatainen, eds., Academic Press, London, 1973, p. 61.

70. Z. Sery, E. Keprt, and M. Obrucknik, J. Thorac. Cardiovasc. Surg., 57, 549 (1969).

71. M. S. Dunnill, Thorax, 17, 329 (1962).

72. J. E. Emery and P. F. Wilcox, Acta Anat., 65, 10 (1966).

73. Y. Kikkawa, E. Motoyama, and L. Gluck, Amer. J. Path., 52, 177 (1968).

74. C. G. Loosli and E. L. Potter, Amer. Rev. Resp. Dis., 80, 5 (1959).

75. M. E. Avery and B. D. Fletcher, The Lung and Its Disorders in the Newborn Infant, 3rd ed., W. B. Saunders, Philadelphia, 1974, p. 3.

76. S. L. Kauffman, P. H. Burri, and E. R. Weibel, Anat. Rec., 180, 357 (1970).

77. T. Crocker, A. Teeter, and B. Nielsen, Cancer Res., 30, 357 (1970).

78. D. Bartlett, Jr., Resp. Physiol., 9, 58 (1970).

79. P. H. Burri and E. R. Weibel, Resp. Physiol., 11, 247 (1971).

80. J. S. Brody and W. J. Buhain, Resp. Physiol., 19, 344 (1973).

81. B. Wu, Y. Kikkawa, M. M. Orzalesi, E. K. Motoyama, M. Kaibara, C. J. Zigas, and C. D. Cook, Biol. Neonate, 22, 161 (1973).

82. E. L. Cunningham, J. S. Brody, and B. P. Jain, J. Appl. Physiol., 37, 362 (1974).

83. D. Bartlett, Jr., Resp. Physiol., 9, 50 (1970).

84. L. K. Romanova, E. M. Leikina, and K. K. Antipova, Bull. Exp. Biol. Med., 63, 303 (1967).

85. J. J. Longacre, and R. Johansmann, J. Thorac. Surg., 10, 131 (1940).

86. J. L. Bremer, J. Thorac. Surg., 6, 336 (1937).

87. W. J. Buhain and J. S. Brody, J. Appl. Physiol., 35, 898 (1973).

88. S. Sorokin, Develop. Biol., 3, 60 (1961).

89. T. Alescio and A. Cassini, J. Exp. Zool., 150, 83 (1962).

90. S. Sorokin in Organogenesis, R. L. DeHaan and H. Urspring, eds., Holt, Rinehart and Winston, New York, 1965, p. 467.

91. L. Reid, in Development of the Lung, A. V. S. deReuck and R. Porter, eds., Little Brown, Boston, 1967, p. 109.

92. A. A. deLorimer, D. F. Tierney, and H. R. Parker, Surg., 62, 12 (1967).

93. K. A. Piez, E. A. Eigner, and M. S. Lewis, Biochemistry, 2, 58 (1963).

94. P. H. Byers, K. H. McKenney, J. R. Lichtenstein, and G. R. Martin, Biochemistry, 13, 5243 (1974).

95. J. R. Lichtenstein, P. H. Byers, B. Smith, and G. R. Martin, Biochemistry, in press (1975).

96. R. G. Crystal, A. W. Nienhuis, N. A. Elson, and W. F. Anderson, J. Biol. Chem., 247, 5357 (1972).

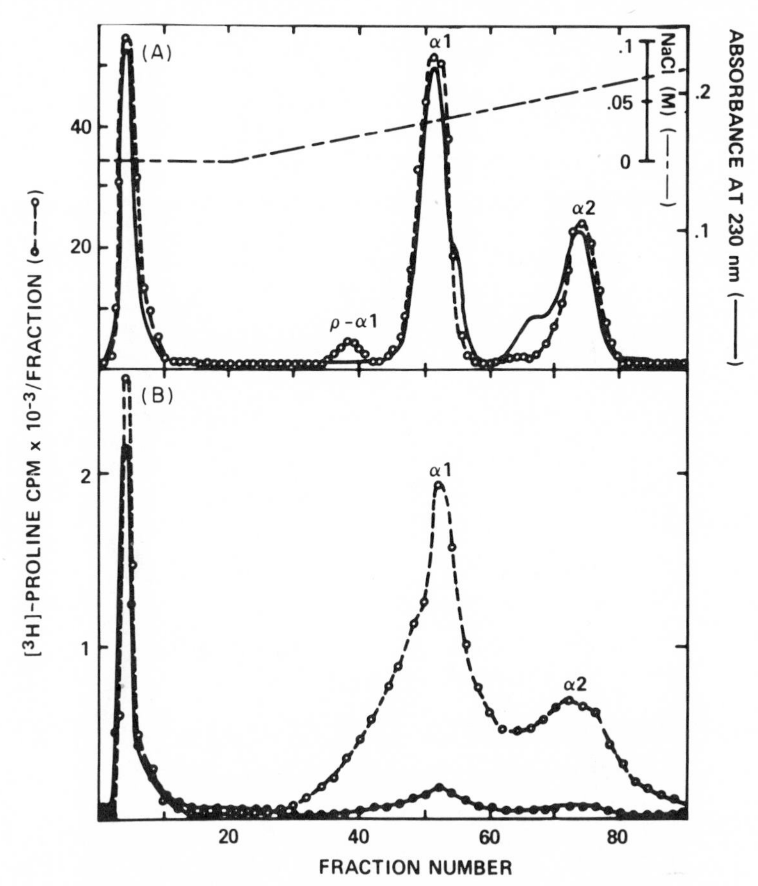


FIG. 1. (A) [$^3$H]proline-labeled collagen, synthesized by explants of two-week-old rabbit lung, was partially purified by acetic acid extraction and subsequently chromatographed on CM-cellulose, with lathyritic rabbit skin collagen as carrier [22].

(B) A cell-free reaction mixture described in Table 3 utilizing fetal-rabbit-lung bound polysomes was scaled to 1 ml, incubated for 60 min at 37° C, and chromatographed on a Sepharose 6B column [25]. The labeled material eluting from the column in the region of collagen chains was pooled, lathyritic skin collagen was added as carrier, and the mixture was chromatographed on CM-cellulose [25]. Counts per minute per fraction (O----O) and per fraction after treatment with collagenase after incubation (●----●). The pattern of carrier collagen (not shown) was identical to (A).

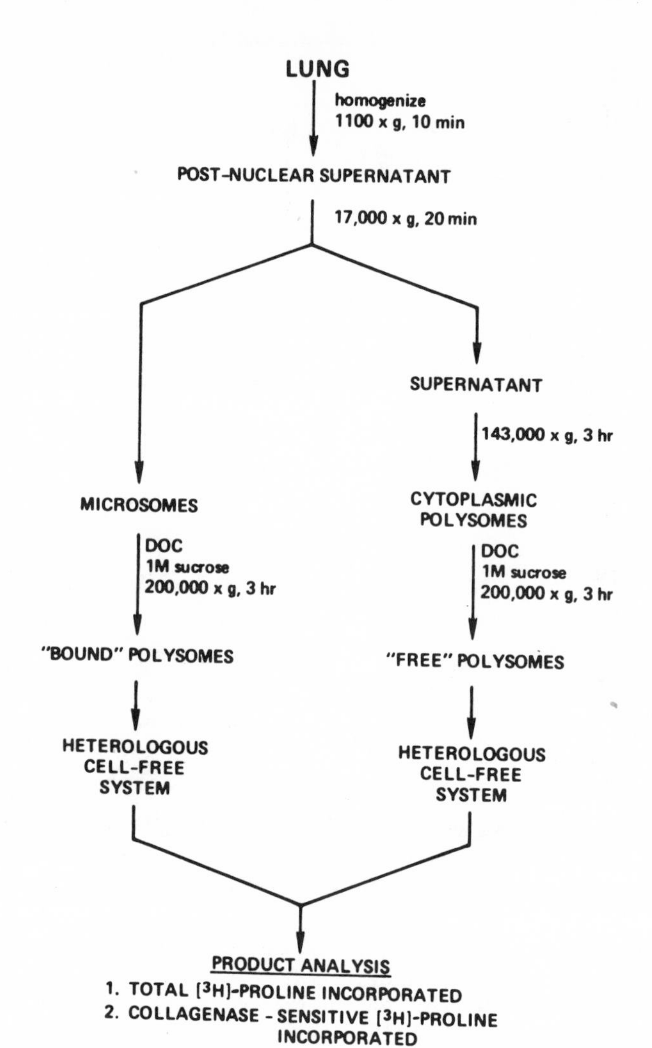


FIG. 2. The lung cell-free system. Polysomes are prepared from rabbit lung homogenized in 0.12M sucrose, 20 mM Tris-HCl (pH 7.4), 240 mM KCl, 7.5 mM $MgCl_2$ [63]. The final polysome preparations are suspended in polysome storage buffer and stored indefinitely in liquid nitrogen vapor [25].

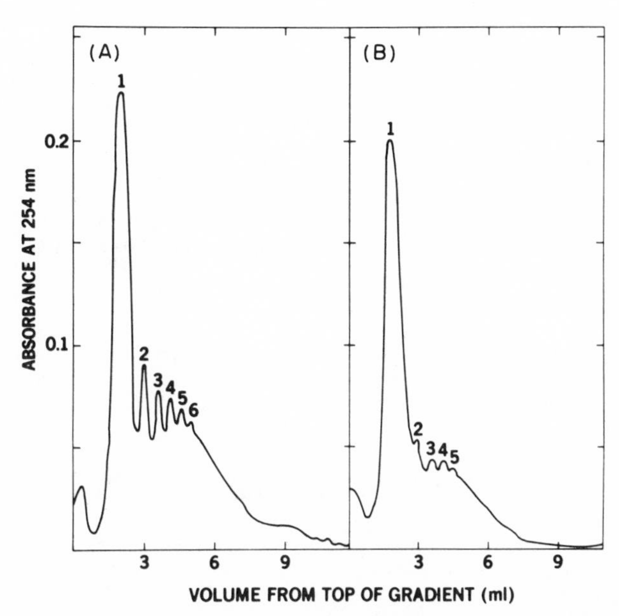


FIG. 3. Lung polysomes from fetal rabbits displayed on 15-35% sucrose gradients. Convex, exponential gradients were prepared, centrifuged, and anlyzed as described elsewhere [25, 96].

(A) 1.9 $A_{260}$ units of bound polysomes.
(B) 2.1 $A_{260}$ units of free polysomes.

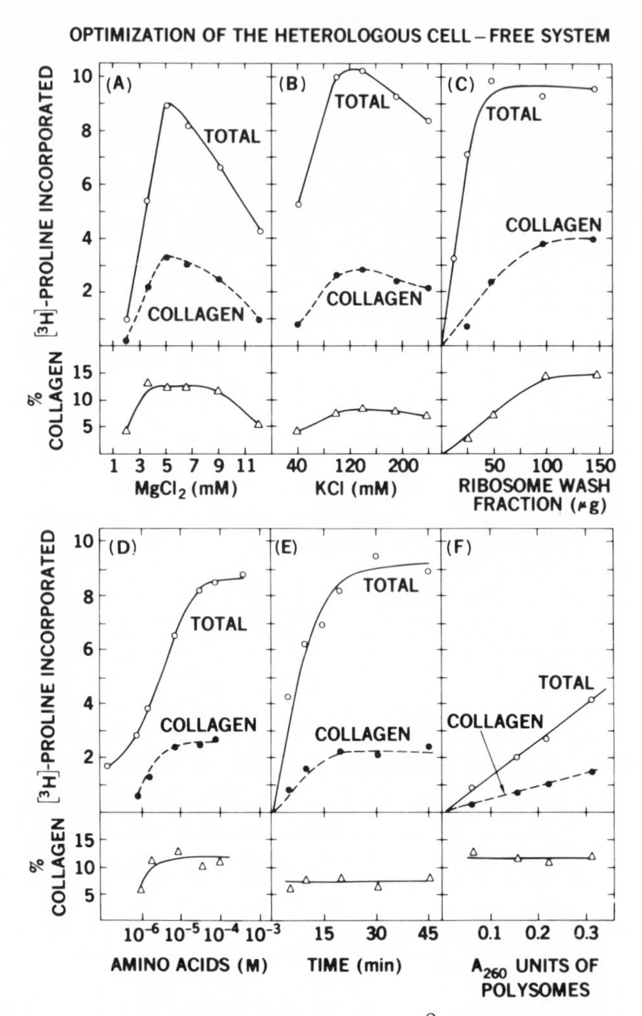


FIG. 4. Incubations (100 μl, 37° C) were carried out varying the $MgCl_2$, KCl, ribosomal wash fraction, amino acid concentration, time of incubation, and amount of polysomes. In panels A-D and F, a 15-min incubation was used. Except in panels in which they were varied, the standard concentrations were 5 mM $MgCl_2$, 123-160 μg of wash fraction, and 40 μM amino acids. Counts per minute of total [$^3$H]proline (O----O) and of collagenase-sensitive [$^3$H]proline (●----●) incorporated per $A_{260}$ unit of polysomes are shown (except panel F, which gives absolute quantities synthesized). The lower segments of each panel show the percent collagen synthesis corrected for the excess proline present in collagen (Δ----Δ).

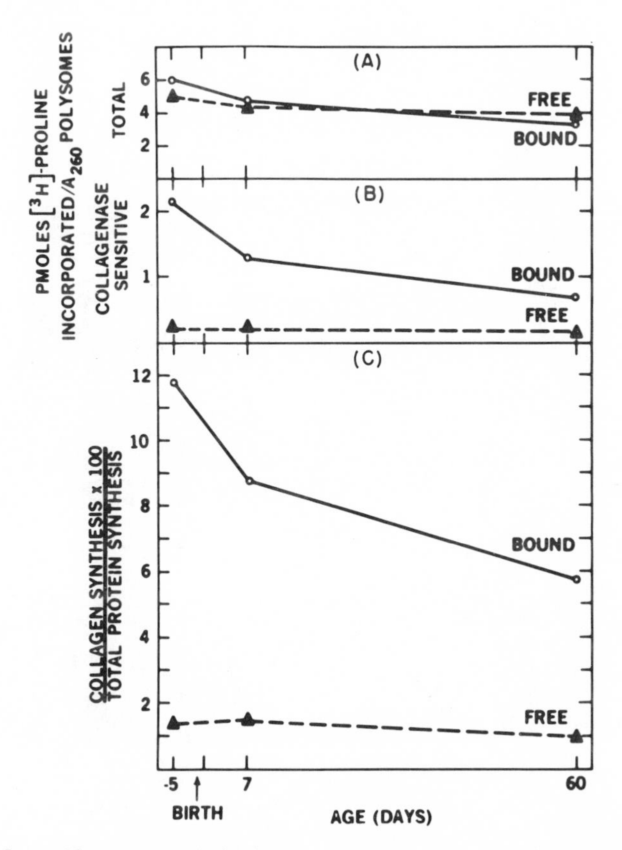


FIG. 5. Changes with lung growth of total protein synthesis, collagen synthesis, and percent collagen synthesis in a heterologous system utilizing lung polysomes. The cell-free system used was identical to that described in Table 3; data with polysomes from the cytoplasm (Δ----Δ) and polysomes derived from the endoplasmic reticulum (0----0) are shown. (A) Total protein synthesis as [$^3$H]proline incorporated per $A_{260}$ unit of polysomes in 15 minutes; (B) collagen synthesis as measured by the collagenase assay (see text); the units are as in (A); (C) percent collagen synthesis, corrected for the excess proline present in lung collagen [22, 25].

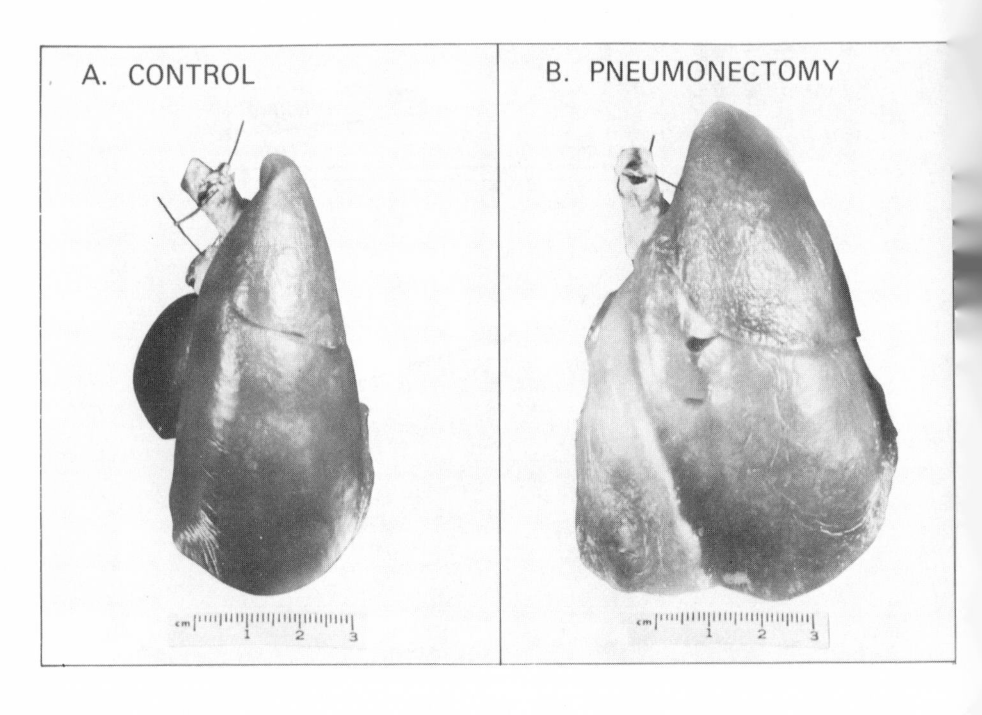


Figure 6. The right lung 28 days after surgery. A left pneumonectomy was performed on a ten-week-old New Zealand white rabbit, and a left thoracotomy on a litter mate control matched for sex and weight [61]. On the 28th postoperative day, the animals were killed and bled. The right lungs were removed and inflated to 15 cm water pressure with buffered formalin (pH 7.4) and allowed to fix for 48 hours. (A) Control; (B) pneumonectomy. No qualitative difference between the lung morphology or connective tissue architecture could be seen at the light microscopic level [61].

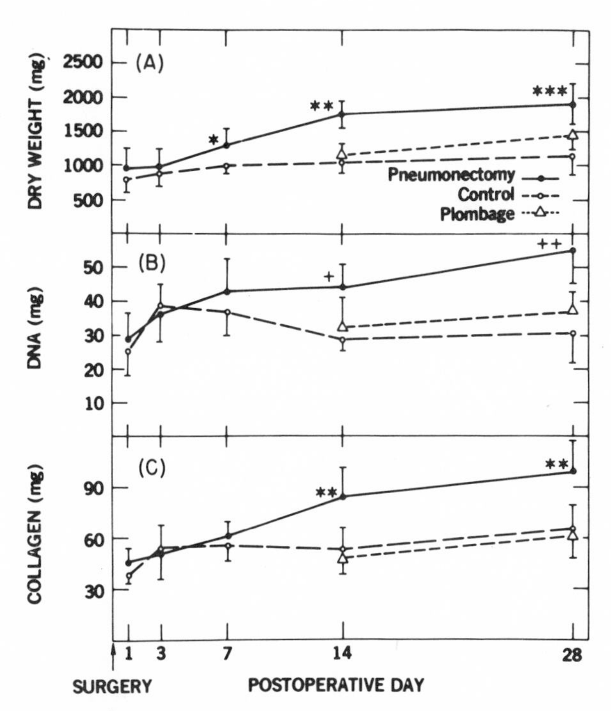


FIG. 7. Changes inthe total amounts of various biochemical parameters of the right lung after left pneumonectomy. A total of 113 animals were studied at 1, 3, 7, 14, and 28 days after pneumonectomy (—●—), control thoracotomy (—O—), or pneumonectomy and wax plombage (--- Δ ---) [61]. The bars at each time point represent one standard deviation. Significance values were calculated by the two-tailed t-test for differences between pneumonectomy and control ($P_{p-c}$), pneumonectomy with plombage ($P_{p-pl}$), and pneumonectomy with plombage and control ($P_{pl-c}$). Significance is represented at the time points by the following:

* $P_{p-c} < 0.005$

** $P_{p-c} < 0.005$, $P_{p-l} < 0.005$, $P_{pl-c} > 0.3$

*** $P_{p-c} < 0.005$, $P_{p-l} < 0.01$, $P_{pl-c} > 0.005$

$P_{p-c} < 0.005$, $P_{p-pl} < 0.05$, $P_{pl-c} > 0.3$

$P_{p-c} < 0.005$, $P_{p-pl} < 0.05$, $P_{pl-c} > 0.05$

Chapter 6

# CHARACTERIZATION OF MYOSIN HEAVY CHAIN mRNA FROM CHICKEN EMBRYONIC MUSCLES

Satyapriya Sarkar
Department of Muscle Research
Boston Biomedical Research Institute
and
Department of Neurology
Harvard Medical School
Boston, Massachusetts

## I. INTRODUCTION

Myosin is one of the major myofibrillar proteins present in all muscle systems thus far studied (for a review, see Ref. 1). The physicochemical properties, as well as the role of this important contractile protein in the contraction-relaxation cycle of muscle tissue, have been studied by several laboratories for many years [1]. A brief discussion of the relation of myosin to muscle structure and function is appropriate so that we may obtain a proper perspective of the roles of messenger RNAs (mRNAs) coding for myosin subunits in processes related to myogenesis and normal contractility in muscle tissue.

The basic units of the contractile system, as shown by studies on striated muscle, are the sarcomeres, and within the sarcomeres are localized the major contractile proteins in regular arrays of interdigitating thick and thin filaments. The thin filaments extend from both ends of the sarcomeres toward the center but do not normally overlap, whereas the thick filaments lie in the centers of the sarcomeres and overlap the thin filaments. The thick filaments consist of aggregates of myosin molecules and the thin filaments are complexes of actin, tropomyosin, and troponin. The actin monomers are spherical units of globular (G-)actin which are arranged into double-stranded helical polymers (F-actin) (for a review see Ref. 2). A myofilament generates force by sliding relative to another. Crossbridges containing the "heads" of the myosin molecules project from the thick filaments and, during contraction, interact with the actin monomers in thin filaments. The two proteins, tropomyosin and troponin, located in thin filaments, are necessary to confer $Ca^{2+}$ sensitivity to actin-myosin interaction and are called regulatory proteins. It is now believed that the "molecular" unit of of contraction-relaxation cycle in muscle consists of seven actins, one tropomyosin, one troponin, and one myosin.

According to our current views, the myosin molecule consists of two large polypeptide chains (heavy chains), each having a molecular weight of about 200,000. The helical segments of the two intertwined heavy chains constitute the rigid rod part of the molecule (the "tail"), leaving the ends of each chain free to form two globular "head" regions. Each head carries one binding site for actin and one site at which ATP is enzymatically hydrolyzed. The head region also contains a number of low molecular weight proteins in the 20,000 dalton range. These proteins (the light chains) are noncovalently attached to the heavy chains. Dissociation of the myosin molecule into heavy and light chains under a variety of denaturing conditions results in a loss of ATPase and actin-binding activity [1]. The myosin monomers have a molecular weight of about 460,000 and are made up of six subunits -- two heavy chains and four light chains. The amino ends of the heavy chains are folded to form the two heads. As shown by limited proteolysis, myosin is cleaved into a helical fragment, light meromyosin (about 150,000) and another fragment, heavy meromyosin (about 350,000), which contains the head region and part of the tail [4]. Further proteolysis of heavy meromyosin releases an α-helical fragment (subfragment-2, molecular weight about 62,000) and two globular individual heads (subfragment-1, 100,000-120,000 daltons). The latter carry actin-combining and ATPase sites. In the myofibril, the light meromyosin fraction of myosin constitutes the backbone of the thick filaments, whereas the cross-bridges contain the heavy meromyosin portion of the molecule [2].

Myosins from different types of muscle cells (e.g., slow, fast, cardiac, smooth) in the same organism, as well as from different organisms, vary considerably in properties. In particular, variation in two important properties of myosins, ATPase activity and self-assembly into filaments, seems to influence the contractile properties of the fibers strongly. Differences in the aggregation properties of myosins are likely to be due to variations in the primary structure of the α-helical light meromyosin portion of the heavy chains, whereas the differences in catalytic properties are due to variations in the globular head region of the heavy chains where the light chains are assembled. These observations support the view that, with increasing physiologic demands for various types of specialized muscle structure and function, a large number of structural genes coding for various heavy and light chains of myosins have evolved. According to this view, different sets of structural genes for myosin subunits are expressed in various types of differentiated muscle fibers (e.g., skeletal, cardiac, smooth), resulting in different isozymic forms of myosin characteristic of each muscle type [5-8]. Recent evidence indicates that there is a minimum of five and possibly more functionally related but different genes for the light chains in fast, slow, or cardiac muscle cells [5,6,8] . Although a large amount of information on the properties of the myosin subunits is available, very little is known about the regulatory mechanisms involved in the biogenesis and assembly of this complex protein.

It is now believed that actin and myosin are present in a large variety of nonmuscle eukaryotic cells (for a review, see Ref. 9). The light chain complement of nonmuscle myosins (also called cytoplasmic myosins) is different from that in muscle myosins [9]. Myosins from nonmuscle cells also exhibit different in vitro aggregation properties as compared to myosins from striated muscles [9, 10]. These differences are apparently due to the formation of smaller and less well-defined filaments. Results indicate that the genes coding for nonmuscle myosin subunits must differ from the corresponding genes expressed in striated muscles. It has been suggested that cytoplasmic actin and myosin may play a role in the processes of cellular motility and contraction in nonmuscle systems [9]. Since these processes are common to almost all eukaryotic cells, the presence of of nonmuscle actin and myosin is now considered a common feature of the biology of many of these cells [9,11]. This consideration obviously has important implications for the genetic regulation of cellular differentiation.

A key problem in myogenesis, which is not yet understood, is the biogenesis of the myofilaments and its regulation. The cell-free system of Heywood and Rich [12] developed from embryonic chicken leg muscles for in vitro synthesis of various myofibrillar proteins [12] is ideally suited for the study of some aspects of this problem in detail. Using this system, we [13] and others [14] have recently shown that the heavy and light chains of myosin are synthesized in vitro on separate polysomes of different sizes. These results clearly indicate that the myosin subunits are translated from monocistronic messages as products of separate genes. Heywood and his group detected a rapidly labeled RNA fraction, 26 to 28S, in the fraction of myosin-synthesizing polysomes by in ovo labeling of embryos with $^{32}PO_4$ [15]. Subsequently, they were able to translate, in vitro, the polysomal RNA containing this 26 to 28S RNA fraction into a product identifiable as the myosin heavy chain [15, 16]. They have also reported that cell-specific initiation factor IF-3 is required for specific recognition of mRNAs (e.g., muscle IF-3 for muscle mRNAs) for successful translation in the heterologous systems they used [16,17]. Previous studies on cell-free translation of myosin heavy chain mRNA have used RNA preparations in which ribosomal RNAs (rRNAs) constituted the bulk of the material [15-17]. In order to study the translation and assembly of different subunits of myosin, as well as the regulatory mechanisms involved in mRNA metabolism during muscle cell growth and differentiation, we have attempted to purify heavy chain mRNA. In this chapter are outlined studies on the characterization of highly purified heavy chain mRNA isolated from chicken embryonic muscles. Although the specific preparative procedures used in these studies are described, no attempt is made to review the literature dealing with eukaryotic mRNA isolation, since there are excellent recent reviews on the subject [18,19].

## II. SOURCES

Homogenates of chicken embryonic leg muscles, which in general have a higher protein synthetic activity than adult muscles, have been used [15-17, 21] for the preparation of myosin-synthesizing polysomes and heavy chain mRNA; mRNA preparations have also been reported from cultured muscle cells [22, 23]. The preparative procedures in general include the following steps: Separation of myosin polysomes from polysomes of smaller sizes, isolation of total polysomal RNA, and separation of mRNA from rRNAs. Since heavy chain mRNA sediments very near 28S rRNA in sucrose gradients [15], the total RNA pooled from the 28S rRNA region in sucrose gradients has also been used in several studies as the source of the mRNA for in vitro translation [15-17] or for further purification [22].

## III. PREPARATION OF HEAVY CHAIN mRNA

The usual source is the myosin polysomes that are obtained by fractionation of muscle polysomes by sucrose gradient centrifugation [15,20,21]. Alternatively, the 28S rRNA region of sucrose gradient runs of total polysomal RNA may be used as the source of the mRNA [15-17]. The procedure developed in our laboratory for obtaining highly purified mRNA preparations in quantities suitable for biochemical studies is summarized in Figure 1. The essential features of the method are: fractionation of polysomes according to size to obtain undergraded preparation of myosin polysomes and separation of mRNA from rRNAs utilizing the unique properties of the poly(A) segment present in the mRNA.

A. Isolation of Myosin-Synthesizing Polysomes

For large scale preparation of polysomes our previously described procedure [13,20] is modified as follows: Thigh muscles of 14-day-old chicken embryos were dissected free of skin and bone under chilled buffer A, which consists of 0.01M Tris-HCl (pH 7.6), 0.25M KCl, 0.01M $MgCl_2$, 0.25M sucrose. (Schwarz/Mann, RNase free), 0.1 mM DTT, 0.1 mM EDTA, and 500 μg/ml of heparin. About 250 g of muscle is obtained from 300 embryos. Ten gram batches of tissues in an equal volume of buffer A were gently homogenized with five strokes in a loose-fitting Dounce homogenizer (Blaessig Glass, Rochester, N. Y.; clearance 0.1 mm).

The homogenate was centrifuged at 12,000 g for 10 minutes (10,000 rpm in a Sorvall SS-34 rotor) to remove the nuclei, unbroken cells, mitochondria, and cell debris. The pellet was washed once by gently resuspending in 0.3 volumes of buffer B (the same as buffer A, except sucrose is omitted), followed by centrifugation at 12,000 g for ten minutes. The supernatants were combined and passed through three layers of sterile cheesecloth to remove fatty material. The cytoplasmic extract thus obtained was then centrifuged at 100,000 g (37,000 rpm in a Beckman Ti 60 rotor) for one hour to pellet the polysomes. The polysomes were gently resuspended in buffer B to a concentration of about 200-250 $A_{260}$ units/ml, and 4 to 5 ml of the polysome suspension was layered on a discontinuous sucrose gradient containing four 13-ml layers: 35%, 30%, 25%, and 20% sucrose (w/v) in buffer B. The gradients were centrifuged at 76,900 g for 120 minutes in a Spinco SW 25.2 rotor (25,000 rpm) and were separated into 1-ml fractions. The bottom ten fractions, together with a loose pellet that sedimented at the bottom of the tubes, were pooled to obtain a polysomal fraction partially enriched in myosin polysomes (for details of sucrose gradient profiles of chicken embryonic leg muscle polysomes,see Fig. 2 in Ref. 13). The pooled fractions were diluted with 3.5 volumes of buffer B and centrifuged at 160,000 g for 120 minutes (48,000 rpm in a Beckman Ti 60 rotor) to pellet the polysomes. The pellet was gently resuspended in buffer B to a concentration of 60-70 $A_{260}$ units/ml; then 1.5 ml of the polysome suspension was layered on a 28-ml 15 to 40% linear sucrose gradient in buffer B. After centrifugation at 64,700 g for 120 minutes in a Spinco SW 25.1 rotor (25,000 rpm), the gradients were fractionated in an ISCO gradient fractionator and the fractions were monitored for absorbance at

260 nm. The bottom six fractions, as indicated by the bar in Figure 2, representing the peak of myosin polysomes, were pooled, diluted with 3.5 volumes of buffer B, and centrifuged at 160,000 g for 90 minutes to obtain a polysomal pellet. All operations were performed at $4^{\circ}$ C. About 75 $A_{260}$ units of purified polysomes are obtained by this procedure from 300 embryos.

### B. Isolation of Polysomal RNA from Myosin-Synthesizing Polysomes

The polysomal pellets obtained from 20-25 gradients were gently suspended in 1% SDS dissolved in buffer C (0.1M Tris-HCl, pH adjusted to 9.0 at $20^{\circ}$ C, 0.1M NaCl, and 1 mM EDTA) to give a concentration of 25 $A_{260}$ units/ml. An equal volume of phenol-chloroform-isoamyl alcohol (50:50:1) was added and the mixture was shaken vigorously for 20 minutes at room temperature [24]. The mixture was then chilled to $5^{\circ}$ C, and the phases were separated by centrifugation at 12,000 g for 20 minutes at $4^{\circ}$ C. The upper aqueous layer was removed and the interphase plus phenol phase was reextracted again with phenol-chloroform-isoamyl-alcohol as described above. The final aqueous phase was made 2% in potassium acetate (pH 5.5). Two volumes of 95% EtOH, chilled to $-20^{\circ}$ C, were added and the mixture was kept at $-20^{\circ}$ C for

12 to 16 hours. The RNA was collected by centrifugation at 12,000 g for 20 minutes. The RNA pellet was washed twice with 66% EtOH containing 0.1M NaCl and dissolved in buffer D (0.01M Tris-HCl, pH 7.6, 0.1M NaCl and 2 mM EDTA) at a concentration of 100 - $A_{260}$ units/ml. The RNA was then precipitated again with ethanol as described above to remove all the traces of dodecyl sulfate. The RNA was dissolved in buffer E (0.01M Tris-HCl, pH 7.6, 0.5M KCl, 1 mM $MgCl_2$) for binding to Millipore filters. About 50- $A_{260}$ units of polysomal RNA was obtained by this procedure from 250 g of leg muscles.

C. Millipore Filtration of Polysomal RNA

It is currently believed that almost all eukaryotic mRNAs, with the exception of histone mRNA, contain ribopolyadenylate [poly (A)] segments of about 150-200 nucleotides, covalently linked at the 3' end [18, 25]. The procedure used by us for the separation of an mRNA from rRNAs is based on some of the unique properties of the poly (A) segments of mRNAs, such as binding to Millipore filters [26] and cellulose columns [27]. The rRNAs that do not contain poly(A) segments preferentially do not bind to the filters or the cellulose column, whereas the mRNAs that contain large poly (A) segments remain strongly bound [26,27].

In our procedure, 75 to 100 $A_{260}$ units of crude polysomal RNA was diluted with buffer E to 0.2 $A_{260}$ unit/ml. The RNA solutions were passed slowly (no pressure applied) at $4^{o}$ C through a Millipore filter (type HAWP, 24 mm, 0.45 μm, Millipore Corp.) previously equilibrated with buffer E, at a rate of 0.4-0.5 ml/min. The filters were washed five times with 10-ml portions of the same buffer. The adsorbed RNA was eluted from the filters as described below:

The filters in batches of three to four were gently agitated for two minutes with 2 ml of buffer E in a sterile Petri dish. The buffer was removed and discarded. The filters were then cut into small pieces with sterile scissors and were kept in 1 ml of buffer F (0.01M EDTA, titrated to pH 5.0 with 1M NaOH and containing 0.5% SDS) for 30 minutes at $4^{o}$ C with occasional gentle shaking. The buffer was replaced and the elution was repeated. The combined eluate fractions were chilled to $0^{o}$ C and centrifuged at 5,000 g for 10 minutes to remove crystals of potassium dodecyl sulfate (these crystals form because of the residual potassium chloride retained on the filters, but they do not interfere with the elution of Millipore-bound RNA). The crystals were washed once with 0.5 ml of an ice-cold solution of buffer F followed by a second centrifugation, as described above, and the washings were combined with the eluate. The eluate fractions were adjusted to a final concentration of 0.1M NaCl and titrated to pH 7.0 by the addition of solid Tris-base. The RNA was then precipitated with two volumes of ethanol. The ethanol precipitation was repeated three times to remove all traces of SDS [20]. About 2.5 to 3 $A_{260}$ units of Millipore-bound RNA were obtained by this procedure from 300 embryos.

### D. Chromatography of Millipore-bound RNA on Cellulose Columns

The RNA samples bound to and eluted from the Millipore filters were chromatographed on columns of cellulose [27] to remove the residual rRNA that was still present in the Millipore-bound fractions. As in the case of Millipore filtration, the separation of rRNA and poly(A)-containing mRNA by chromatography on cellulose columns depends on the property of the poly(A) segments to adsorb strongly to cellulose [27]. The columns were first calibrated with samples of rRNA and poly(A) run as standards. The rRNA prepared from chicken embryonic leg muscle monosomes (for details of ribosome profiles in sucrose gradient runs see Figure 2 in Ref. 13) and a sample of poly(A) containing an average of 200 adenylic acid residues (Miles Laboratories) were dissolved in buffer G (0.01M Tris-HCl, pH 7.6, 0.5M KCl, and 2 mM $MgCl_2$) at a concentration of 15-20 $A_{260}$ units/ml. Two to three $A_{260}$ units of these ribopolynucleotide samples were applied to two 5.0 x 0.5 cm columns of cellulose (Sigmacell type, 38, microcrystalline, particle size 38 μm, Sigma Chemical Co.) previously equilibrated with buffer G. The poly(A) sample was quantitatively adsorbed on the column (Fig. 3). When the eluant was switched to the low-salt buffer H (2 mM Tris-HCl, 1 mM EDTA, pH 7.0), the poly(A) was eluted as a sharp peak (Fig. 3). The rRNA sample, in contrast, was not retained on the column and was eluted in the wash fraction as a sharp single peak (Fig. 3 A). After the chromatographic procedures for the separation of rRNAs and poly(A)-rich RNA species had been calibrated, the Millipore-bound heavy chain mRNA samples were then further purified on these columns. Five to six $A_{260}$ units of RNA, combined from two to three batches of Millipore filtration,were applied to the column and chromatography was performed as described above. Two well-resolved peaks were obtained: an unadsorbed early peak, which was eluted in the wash fraction like rRNA, and a second peak of adsorbed RNA, which was eluted like poly-(A)-rich RNA with buffer of low ionic strength (Fig. 3 B).

The relative amount of RNA in the early peak varied in different chromatographic runs and ranged from 30 to 75% of the total UV-absorbing material. This variation presumably depends on the efficiency of partition of mRNA and rRNAs on Millipore filters. The RNA fractions pooled from the two peaks were concentrated by precipitation with ethanol and sodium chloride, as previously described.

To prevent nucleolytic degradation, acid-washed glassware, RNase-free sucrose, sterile pipettes, and heparin--an RNase inhibitor--were routinely used. Buffers and reagents were made in freshly autoclaved water, and all operations were performed at 4 $^{o}$C, unless otherwise specified.

### E. Polyacrylamide Gel Electrophoresis of RNA Fractions Purified by Cellulose Column Chromatography

The RNA fractions obtained from the two peaks from cellulose column chromatography were analyzed by polyacrylamide gel electrophoresis using 2.5% gels cross-linked with 0.175% Bis [28]. The RNA samples were dissolved in a buffer containing 30 mM $NaH_2PO_4$, 36 mM Tris, and 1 mM $Na_2EDTA$, (pH 7.8), 0.2% SDS, and 10% sucrose. After the gels were run for 30 minutes at 5 mA/tube, the RNA samples (15 to 30 μg) were applied to the gels, and electrophoresis was resumed at 5 mA/tube. When electrophoresis was complete, the gels were removed and washed in the electrophoresis buffer for 3 hours to reduce the background density in the gel [28]. The gels were scanned at 260 nm in a Gilford recording spectrophotometer fitted with a linear transport device for scanning gels.

As shown by the densitometric scans of the gel runs, the early unadsorbed RNA peak consists of 28S rRNA as the major species and a small amount of 18S rRNA (Figure 4, dotted line). This is consistent with the observation that among the two species of rRNAs, 18S rRNA shows less binding to the Millipore filters [26]. The adsorbed RNA peak from the cellulose column whose chromatographic properties resemble those of a poly(A)-rich mRNA (Figure 3) gave a sharp single peak in the gel scan moving slower than 28S rRNA (Figure 4, solid line). No detectable amounts of other cellular RNA species were present in this RNA sample. The sensitivity of the method would have made it possible to detect other contaminating RNA species present in quantities as low as 2-3% of the RNA used in the gel run. These results indicate that by using the combination of Millipore filtration and cellulose column chromatography a homogeneous preparation of a nonribosomal, poly(A)-rich RNA species present in myosin polysomes can be obtained. (Occasionally, the gel scans indicate the presence of a small shoulder of UV-absorbing material, presumably due to trace amounts of residual 28S rRNA in the mRNA-like fraction. This shoulder can be removed by rechromatography of the RNA sample on a second cellulose column.)

Although the procedure originally developed in our laboratory is based on sequential application of Millipore filtration and cellulose column chromatography, other chromatographic methods, such as the use of oligo(dT)-cellulose after Millipore filtration, also give the same degree of purification of heavy chain mRNA.

## IV. IN VITRO TRANSLATION OF MYOSIN HEAVY CHAIN mRNA

The capacities of the mRNA samples prepared by the methods outlined above for translation of myosin heavy chain were tested using two types of in vitro protein synthetic systems, a homologous cell-free system consisting of salt-washed embryonic chicken leg muscle ribosomes and unfractionated muscle initiation factors [15,16,20], and a heterologous rabbit reticulocyte lysate [29]. The products of translation of the mRNA were subjected to a number of purification steps previously used by us [13] and others [12,14] to demonstrate in vitro synthesis of the myosin heavy chain. These include coprecipitation at low ionic strength with highly purified chicken leg myosin added as carrier after removal of excess radioactive amino acids by dialysis, chromatography of the myosin heavy chain on DEAE-cellulose, and SDS-polyacrylamide gel electrophoresis of the chromatographed myosin. The myosin heavy chains share many physicochemical properties of the heteropolymeric myosin molecule (i.e., the protein consisting of both heavy and light chains). These include insolubility at low ionic strength, presumably due to the intertwined α-helical light meromyosin segment of the heavy chains, and characteristic elution properties upon chromatography on DEAE-cellulose. These latter properties are possibly due to both heavy and light meromyosin portions of the heavy chains. Details of these purification steps and the in vitro translation of heavy chain mRNA in the homologous system are available in a recent publication from this laboratory [20]. I will, therefore, attempt to outline here the successful translation of this mRNA in rabbit reticulocyte lysate.

Incubation mixtures used for translation contained in a total volume of 0.5 ml: 0.4 ml of freshly prepared rabbit reticulocyte lysate [29]; 10 mM Tris-HCl (pH 7.4); 2 mM $MgAc_2$; 2 mM dithiothreitol (DTT); 1 mM ATP; 0.2 mM GTP; 15 mM creatine phosphate; 30 μg creatine kinase; 150 mM kCl; 50 μM hemin; 10 μM each of 20 different amino acids; 5 μCi of [$^{14}$C]amino acid mixture (New England Nuclear, Inc.); and 5-10 μg of mRNA samples. After incubation at 30° C for 60 minutes, reaction mixtures from ten incubations were pooled, and the KCl concentration was adjusted to 0.5M. Highly purified chicken embryonic leg myosin (5 mg, Ref. 30) was added as a carrier and the sample was dialyzed against 6 liters of a buffer containing 50 mM Tris-HCl (pH 7.5), 0.5M KCl, 1 mM EDTA, and 0.2 mM phenylmethyl sulfonyl fluoride (used as a protease inhibitor). The dialyzed sample was centrifuged at 160,000 g for 120 minutes to pellet the ribosomes. The supernatant was then concentrated to 1 ml with Aquacide (Cal Biochem.) and dialyzed for two to three hours against the same buffer. Myosin was then precipitated by adding nine volumes of cold solution consisting of 1 mM Tris-HCl (pH 7.0) and 1 mM EDTA, and the sample was left at 0° C for 10 to 12 hours. The myosin was collected by centrifugation at 12,000 g for 15 minutes and dissolved in 1 ml of 0.03M $K_4P_2O_7$ (pH 7.5), 0.1 mM EDTA, and 0.1 mM DTT, which was added after fraction nine was collected (Figure 5). The fractions were monitored for absorbance at 280 nm and radioactivity. Two protein peaks were obtained (Figure 5); a peak of nonmyosin proteins which was not adsorbed on the column and was eluted in the wash fraction, and a second peak of adsorbed myosin which was released when the eluant was changed to a buffer of high ionic strength. The radioactivity profiles indicated that about 40% of the total radioactivity applied to the column was eluted with the early, nonmyosin protein peak. This is presumably due to the globin and non-globin polypeptide chains which are synthesized in the reticulocyte lysate and coprecipitated with myosin at low ionic strength. The column fractions corresponding to the myosin peak and indicated by the bar were pooled and concentrated by Aquacide for SDS-gel electrophoresis.

The products of in vitro translation of heavy chain mRNA, purified by DEAE-cellulose chromatography as described above, were further analyzed by SDS-polyacrylamide gel electrophoresis. Myosin was dialyzed against 0.1M Tris-Ac (pH 8.0), 1% SDS, 1% β-mercaptoethanol, and 20 mM DTT. Portions of the myosin samples containing 1000 to 1200 cpm (specific activity 5000 - 5500 cpm/mg) were applied to 8-cm 4% gels and the gels were run at 5 mA/tube for 100 to 120 minutes. The gels were stained with Coomassie Brilliant Blue and then sliced. The slices were digested with hydrogen peroxide and counted [13]. The results, presented in Figure 6, indicate that about 70% of the radioactivity applied to the gel migrated with the 200,000 dalton heavy chain band. No significant amount of radioactivity above the background level was detected in the faster moving light chain (average MW, 20,000) regions of the gels (for details of SDS gel electrophoretic separation of heavy and light chains of myosin see Refs. 5, 13). This result supports our previous conclusion that the heavy and light chains of myosin are synthesized separately on different classes of polysomes programmed by monocistronic messages [13]. To test whether the reticulocyte lysate alone synthesizes any myosin heavy chain-like polypeptide, an aliquot of the total TCA-insoluble protein, obtained from the lysate incubated without added heavy chain mRNA, was also subjected to electrophoresis. About 80% of the radioactivity migrated with the band of hemoglobin run as a marker, and no significant amount of radioactivity was observed in the myosin heavy chain band (Figure 6). When samples of RNA species obtained from the Millipore filtrate and the unadsorbed early rRNA-like peak from cellulose column chromatography were tested for myosin heavy chain synthesis in the lysate, only about 1 to 2% of the radioactivity incorporated coprecipitated with the carrier myosin at low ionic strenthh. When the products were subsequently chromatographed on DEAE-cellulose, the radioactivity in the myosin peak decreased further to negligible amounts. Therefore, subsequent analysis of the sample by SDS gel electrophoresis was omitted. These results indicate that the mRNA activity in myosin-synthesizing polysomes was almost exclusively associated with the poly-(A)-containing RNA species.

The results presented above strongly suggest that in vitro translation of purified heavy chain mRNA gave the 200,000 dalton myosin heavy chain as the product. However, it is necessary to confirm the fidelity of the translation of the mRNA. Thus, the synthesized polypeptides were cleaved with CNBr and the resulting peptides were analyzed for radioactivity and UV absorbance. The products from pooled incubation mixtures were copurified with carrier heavy chains [5,7] by precipitation at low ionic strength and DEAE-cellulose chromatography, as described above. The heavy chains were prepared from chicken leg muscle myosin by treatment with 8M urea or at pH 11 [5,7,31]. With SDS-polyacrylamide gel electrophoresis of the reisolated heavy chains, the absence of light chains was demonstrated. Details of these procedures are described elsewhere [5,7,31]. The in vitro synthesized polypeptides purified with the heavy chains were then aminoethylated [32] and cleaved with CNBr [33]. The resulting peptides, after removal of excess CNBr, were dissolved in 2 ml of 70% HCOOH and were separated according to size by molecular sieving on two columns (70 cm x 2 cm) connected in series of Sephadex G-50 equilibrated with 25% HAc [33]. Peptides were eluted with 25% HAc and fractions were monitored for absorbance at 280 nm and radioactivity. Details of the experimental procedures used for the CNBr treatment and processing of the resulting peptides used in these studies are modeled on the studies of Elzinga on CNBr-peptides of rabbit skeletal actin [33]. The results, presented in Figure 7, indicate that 12 UV-absorbing peaks could be separated on these columns. Some of the peaks represent mixtures of peptides that are incompletely resolved and that may be separated into a larger number of resolved peptides by further chromatography (for a detailed discussion of a similar analysis of CNBr-peptides of actin see Ref. 33). Each radioactive peak from the in vitro products corresponds to a UV-absorbing peak. The fractions that did not absorb UV light did not contain any radioactivity. The total amount of radioactivity recovered in the peptides amounted to about 90% of the input. These results confirm that the mRNA indeed directs the de novo synthesis of myosin heavy chains.

It may be appropriate to comment about the requirement for mRNA-specific factors reported in the literature for the successful translation of myosin heavy chain mRNA in cell-free systems [16,17,34]. Heywood and his group have reported that embryonic chicken leg muscles contain a multiplicity of initiation factors (IF-3) that are required for specific recognition of various mRNAs, such as myosin heavy chain mRNA and myoglobin mRNA, when assayed in heterologous cell-free systems (for a recent review on the initiation of protein synthesis, see Ref. 35). More recently, they have reported that a low molecular weight RNA species, isolated from IF-3 fractions, specifically inhibits the translation of heterologous mRNAs [34]. The authors have postulated that this RNA, in conjunction with tissue-specific IF-3, is responsible for mRNA discrimination -- a mode by which protein synthesis may be regulated during differentiation of eukaryotic cells. Several recent reports of experiments utilizing globin and viral mRNAs have also suggested that mRNA-specific initiation factors are found in some eukaryotic cells [36, 37]. Studies by many workers, on the other hand, have conclusively shown that various mRNAs from highly differentiated cells (e.g., lens -- α-crystalline mRNA) can be translated faithfully in cell-free systems derived from totally different tissues without any factor requirements [19]. Furthermore, mRNAs from widely different species (e.g., globin and lens α-crystalline mRNAs) are translated with high efficiency when injected into oocytes of Xenopus laevis [38,39] . The oocyte system is believed to mimic the in vivo conditions. In our hands, highly purified heavy chain mRNA has been successfully translated in rabbit reticulocyte lysate without the addition of muscle factors [21, 40] . Although we cannot rule out that discrimination of mRNAs requires optimization of conditions for in vitro translation of each message, and such factors may play a regulatory role in vivo, it is the view of this author that any specific requirements for such factors do not appear to be stringent.

## V. SIZE AND POLYADENYLIC ACID CONTENT OF MYOSIN HEAVY CHAIN mRNA

Two approaches have been used in our laboratory to estimate the size of heavy chain mRNA. These are (a) polyacrylamide gel electrophoresis in the presence of formamide, which results in unfolding of RNA molecules [41], and (b) determination of the size of the polyadenylic acid fragment located at the 3'-end of the mRNA. The size of the poly(A) fragment is then used to calculate the molecular weight of the mRNA. The results obtained by these two methods are presented below.

### A. Polyacrylamide Gel Electrophoresis of mRNA

Highly purified heavy chain mRNA prepared by the methods described (Sects. III C, and III D) was analyzed by polyacrylamide gel electrophoresis in the presence of formamide according to the method of Staynov et al. [41]. Chicken 28S and 18S and *E. coli* 23S and 16S rRNAs were used as markers. Electrophoresis was performed on 2.5% gels and the gels were scanned at 260 nm. A linear relationship between the electrophoretic mobilities and the log of molecular weight of the purified mRNA was obtained (Fig. 8). From this plot, the molecular weight of the purified mRNA was calculated as about $2.23 \times 10^6$--an estimate that corresponds to a ribopolynucleotide of about 6500 nucleotides [21, 40].

### B. Polyadenylic Acid Content of Myosin Heavy Chain mRNA

The approach used in this study is based on selective labeling of the 3'-end of the poly(A) segment with $[^3H]NaBH_4$ after oxidation of the mRNA with $NaIO_4$ [42]. This treatment oxidizes the two hydroxyl groups on ribose carbon atoms 2 and 3 of the terminal 3'-nucleotide to a dialdehyde. Subsequent reduction with $[^3H]NaBH_4$ results in selective tritiation of these carbon atoms to give an mRNA labeled at its 3'-end. The size of the labeled poly(A) segment after cleavage by RNase of the mRNA and isolation was estimated by electrophoresis. Details of the studies have been published [21,40] and only a brief outline is presented here. A prerequisite for using this approach is that the label must be restricted to the 3'-terminus of the mRNA. This was experimentally tested by comparing the radioactivity in the poly(A) fragment of the tritiated mRNA with that of the intact mRNA. When the RNase-treated sample of labeled heavy chain mRNA was passed through a Millipore filter to bind the poly(A) fragment, about 97% of the label was found in the Millipore-bound poly(A) (Table 1). The bound poly(A) fragment was then eluted from the filters with SDS and was precipitated with ethanol after the addition of 4S RNA as carrier. Treatment with SDS released about 82% of the bound radioactivity, 91% of which was recovered after precipitation with ethanol. These results indicate that the label was restricted to the 3'-terminus of heavy chain mRNA.

The isolated tritiated poly(A) fragment of heavy chain mRNA was subjected to polyacrylamide gel electrophoresis according to the procedure of Loening [28], using 7S, 5S and 4S RNAs as markers. After electrophoresis was completed the gels were stained with toluidine blue and scanned to locate the position of the markers. The radioactivity profile due to the labeled poly(A) fragment was determined by slicing the gel and counting the slices. As shown by the densitometric scanning and radioactivity profile, the labeled poly(A) fragment of mRNA migrated between 7S and 5S RNAs as a distinct sharp peak (Figure 9 B), indicating that the poly(A) fragments of heavy chain mRNA were homogeneous in size. Previous studies have shown that estimates of the size of poly(A) fragments of rapidly labeled mRNAs of HeLa cells, calculated from electrophoretic mobilities, are in good agreement with those obtained from in vivo labeling techniques [43]. From a plot of log molecular weight versus distance migrated, the size of the poly(A) fragment was estimated to be $5.7 \times 10^4$ daltons

TABLE 1

Labeling of the Poly(A) Fragment of Myosin Heavy Chain mRNA

| Condition | Total $^3$H Radioactivity in Sample |
|---|---|
| 1. Intact mRNA | 120,900 |
| 2. Millipore-bound poly(A) after RNase treatment of labeled mRNA | 116,000 |
| 3. Poly(A) extracted from Millipore filters | 95,400 |
| 4. Poly(A) precipitated with ethanol | |

[a]Eighty μg of purified mRNA was tritiated [42,50] with 25 mCi of $^3$H $NaBH_4$ (specific activity, 2.8 Ci/mmole; New England Nuclear Corp.). The tritiated RNA was dissolved in 1 ml of 20 mM Tris-HCl (pH 7.5) containing 1 mM EDTA. Key: (1) A portion (0.1 ml) was mixed with 20 volumes of buffer E and filtered through a Millipore filter [26]. (2) After incubation with 2 μg/ml of RNase at 35 $^\circ$C for 30 min., a 0.2-ml aliquot of the remaining sample was diluted with 1 ml of buffer E containing two $A_{260}$ units of 4S RNA and was filtered through a second Millipore filter. (3) The remaining sample was passed through a third filter and the adsorbed poly(A) fragment was eluted twice with buffer F. (4) The poly(A) fragment eluted from the filters plus 3 $A_{260}$ units of 4S RNA, added as carrier, were precipitated with ethanol. The filters were washed, dried, and counted. For details, see text. (Reprinted from Ref. 21, p. 992, by courtesy of Academic Press, Inc.)

(Figure 9 A). This value corresponds to a ribopolynucleotide of 170 adenylic acid residues [21].

After the size of the poly(A) fragment of highly purified heavy chain mRNA was determined, the mRNA was isolated and purified from embryos that had been pulse-labeled in ovo with $^{32}PO_4$. The $^{32}P$-labeled poly(A) fragment was then isolated from the mRNA, and the amounts of radioactivity in the poly(A) fragment and the intact mRNA were compared. As shown in Table 2, the amount of radioactivity in the poly(A) fragment was 2.92% of that in the intact mRNA. Since the estimate represents a polynucleotide fragment of 170 nucleotides, we can calculate that the number of nucleotides in the intact mRNA is about 5830, with a molecular weight of about $2.05 \times 10^6$. This estimate is in good agreement with the value of $2.23 \times 10^6$ obtained by formamide gel electrophoresis.

TABLE 2

Poly(A) Content of $^{32}P$-labeled Myosin Heavy Chain mRNA[a]

| Sample | Total | Radioactivity(%) |
|---|---|---|
| Intact mRNA | 13,400 | 100 |
| Millipore-bound poly(A) fragment after RNase treatment of $^{32}P$-labeled mRNA | 372 | 2.92 |

[a]Ten 14-day-old chicken embryos were injected in ovo with 5 mCi each of carrier-free $^{32}PO_4$. Myosin heavy chain mRNA was isolated after incubation of the embryos at 37° C for 2 hr., and the labeled mRNA was diluted with nonradioactive mRNA to a specific activity of 270 cpm/μg. Then 50 μg of mRNA was incubated with RNase, and the poly(A) fragment was adsorbed on a Millipore filter. For details see text and legends to Table 1. (Reprinted from Ref. 21, p. 994, by courtesy of Academic Press, Inc.)

The mRNA coding for the 200,000-dalton myosin heavy chain should contain a minimum of about 5400 nucleotides or a molecular weight of about 1.88 x $10^6$. The estimates of the molecular weight of the heavy chain mRNA obtained by the two approaches described above range from 2.05 to 2.23 x $10^6$. These values, although 10-20% higher than the minimum, are in good agreement with the expected molecular weight of heavy chain mRNA. Based on these results, the possibility must be considered that some "nontranslatable" segment(s) may be present in the mRNA. In addition to the poly(A) fragment located at the 3'-end of most eukaryotic mRNAs, it is possible that such "nontranslatable" segment(s) may also include polynucleotide sequences at the 5'-end.

Previous work has shown that the mRNA fraction present in myosin polysomes sediments slightly faster than 28S rRNA in sucrose gradients [15, 22]. The studies presented above indicate that the heavy chain mRNA is considerably larger than the chick 28S rRNA, which has a molecular weight of 1.6 x $10^6$ [44]. These two properties -- an apparent low S value (26 to 27S) in sucrose gradients and relatively low mobility in polyacrylamide gels -- strongly suggest that the heavy chain mRNA may possess a considerable amount of secondary structure. Whether this property of the mRNA plays any role in the metabolic events during translation and transcription remains to be explored.

Przybala and Strohman have recently reported that heavy chain mRNA obtained from cultured muscle cells does not bind to oligo(dT)-cellulose, and, therefore, lacks a large quantity of poly(A) [23]. It should be noted that in these studies the heavy chain mRNA was neither purified nor biochemically characterized, and the products of in vitro translation of the mRNA were not subjected to rigorous purification steps. Therefore, it is difficult to reconcile these results with those reported here. The use of a commercial preparation of oligo(dT)-cellulose (Collaborative Research, Inc.), different batches of which are known to give variable elution profiles of poly(A)-rich mRNAs, may account for the results obtained by Strohman's group. To test the possibility that chromatography of heavy chain mRNA on oligo (dT)-cellulose instead of cellulose may give a different elution profile, we have chromatographed a sample of total polysomal RNA from myosin polysomes on a column of oligo(dT)-cellulose (Type $T_3$, Collaborative Research, Inc.). As shown in Figure 10, about 4% of the total UV-absorbing material was adsorbed on the column. This fraction was eluted only when the eluant was switched to a buffer of low ionic strength. Samples corresponding to the two peaks, A (rRNA-like) and B (poly(A)-rich mRNA-like) were then tested for the synthesis of myosin heavy chain in rabbit reticulocyte lysate. About 2% of the total mRNA activity was recovered in the RNA fraction which did not bind to oligo(dT)-cellulose (insert, Figure 10). From these results it is concluded that the heavy chain mRNA isolated from developing chick embryonic muscles behaves like other typical poly(A)-containing mRNAs during chromatography on oligo(dT)-cellulose. Furthermore, the mRNA activity is quantitatively recovered in the bound fraction. In agreement with these results, Buckingham et al. [45] and Kaufman and Gross [46] have recently shown that a poly(A)-containing large RNA species, presumably heavy chain mRNA, is present in cultured muscle cells, although the mRNA was not characterized by in vitro translation in these studies. Another possibility that may account for the failure of heavy chain mRNA to bind to oligo(dT)-cellulose, as reported by Przybala and Strohman [23], is that the poly(A) fragment was cleaved during the isolation procedures used by them. Therefore, the conclusion of these authors that the heavy chain mRNA does not contain detectable lengths of poly(A) must be documented by evidence stronger than that provided.

## VI. REVIEW OF OTHER STUDIES ON MYOSIN HEAVY CHAIN mRNA

Heywood's group [22] has described a procedure for the isolation of heavy chain mRNA from myosin polysomes of cultured muscle cells. The procedure includes isolation of myosin polysomes from sucrose density gradients of cell extracts; adsorption of polysomes on Millipore filters and treatment of the bound polysomes with 0.5% SDS, which releases the polysomal RNA in the filtrate; isolation of a 24 to 30S RNA fraction (containing heavy chain mRNA and 28S rRNA) on density gradients, and removal of the 28S rRNA on a Sepharose 2B column. The last step is based on the capacity of the 28S rRNA to be bound to Sepharose 2B at high ionic strength [47], whereas other RNA species such as 18S rRNA, heavy chain mRNA (and other mRNAs, such as globin mRNA) are not bound under these circumstances [22]. The cultured muscle cells used in these studies were pulse-labeled with [$^3$H]uridine and the isolation of the heavy chain mRNA was monitored by the radioactivity in the RNA fraction. However, the details of the characterization and in vitro translation of the mRNA were not reported in these studies [22].

In a recent report on heavy chain mRNA from cultured muscle cells, Przybala and Strohman [23] have used total RNA extracted from the polysomes and the ribonucleoprotein (RNP) particles released by treatmetn of polysomes with potassium chloride and puromycin [48] for the isolation of an RNA fraction with mRNA activity. The mRNA was identified by cell-free translation in a rabbit reticulocyte lysate, as detected by SDS gel electrophoresis of the incubation mixture after addition of unlabeled myosin as carrier. The heavy chain mRNA was found to sediment at 18 to 26S, was associated with a puromycin-dissociated RNA particle sedimenting between 20 and 40S, and appeared predominantly in a fraction that did not bind to an oligo(dT)-cellulose column (see also Sec. V. B). In agreement with our results [21, 40], the heavy chain mRNA in these studies did not show any stringent requirement for muscle initiation factors to be translated in this heterologous system, in contrast with the results of Heywood [16, 17]. The purification and characterization of the heavy chain mRNA was not reported in these studies [22].

Buckingham et al. [45], in a recent study on the synthesis of poly(A)-containing cytoplasmic RNA species in cultured muscle cells, detected a rapidly labeled 26S RNA species by hybridization to poly(U)-

glass fiber filters and gel electrophoresis. This 26S RNA was present in all stages of growth and differentiation of the cultured muscle cells. During the transition from the nondifferentiated myoblast to the differentiated myotube stage (for a detailed discussion on the growth and differentiation of muscle cells in culture see Ref. 49) this 26S RNA was found to be stabilized with a significant increase in its biological half-life. The authors further concluded that the 26S putative mRNA present in the myoblast stage and those in the myotube stage were identical. Since the mRNA was not characterized by in vitro translation in these studies, the conclusion of the authors regarding the identity of the 26S mRNA in all stages of muscle cell growth and differentiation should be documented by stronger and more definitive results than those presented (see also Sect. I for a discussion on the nonidentity of myosin heavy chains from various muscle and nonmuscle cells).

## VII. CONCLUDING REMARKS

The work reported in this chapter describes a procedure for the isolation and characterization of heayy chain mRNA, which is undegraded and electrophoretically homogeneous, in sufficient quantity to estimate its concentration by UV absorbance. Previous preparations of heavy chain mRNA described in the literature [15,17, 23] contained the mRNA in concentrations too low for direct estimation. The mRNA in these reports was detected either by radioactivity due to in vivo labeleng [15] or by in vitro translation of the mRNA fraction [16, 17, 23] present in the preparations, the bulk of which consisted of rRNAs. These are the first techniques of isolation and characterization of highly purified heavy chain mRNA that allow one to prepare this RNA in quantities suitable for biochemical studies. The heavy chain mRNA coding for a 200,000 dalton polypeptide chain is one of the largest mRNAs that has been described and translated in vitro.

The availability of purified preparations of heavy chain mRNA should make it technically feasible to answer a number of important questions related to the regulatory processes involved in myogenesis. These include the problems of the transcription and translation of the mRNA during muscle cell differentiation; the possible existence of pretranscribed heavy chain mRNA in a "nontranslatable" form before terminal differentiation of the muscle cell; the role of the mRNA, if any, in the transition of the myoblast to myotube stages; and the important question of the identity of heavy chain mRNAs present in various stages of muscle cell differentiation. In addition, the properties of the purified mRNA may be used to study the details of a number of cellular events, such as transcription, nuclear processing, and transport to the cytoplasm, that are common to all eukaryotic mRNAs. Another important area where purified mRNAs coding for the myosin subunits may be useful is the study of in vitro translation and assembly of the subunits and the relation of these processes to the overall process of biogenesis of the myofilaments and its control.

## ACKNOWLEDGMENTS

The author is grateful to Drs. John Gergely, James D. Potter, and Henry Paulus for many helpful discussions during the course of this work and to Dr. John H. Collins for his advice on the analysis of CNBr-peptides. I wish to acknowledge the participation of Dr. Hrishikes Mondal in the initial stage of the purification of heavy chain mRNA.

This work was supported by grants from the National Institutes of Health (AM-13238), the Massachusetts Heart Association (No. 1097), the American Heart Association (No. 71-915), and the Muscular Dystrophy Associations of America, Inc. It is a pleasure to acknowledge the skillful technical assistance of Miss Ann Sutton and Miss Ven-Jim Chen.

## REFERENCES

1. E. W. Taylor, Ann. Rev. Biochem., 41, 577 (1972).
2. H. E. Huxley, Science, 164, 1356 (1969).
3. A. Weber and J. M. Murray, Physiol. Rev., 53, 53, 612 (1973).
4. S. Lowey, H. S. Slayter, A. G. Weeds, and H. Baker, J. Mol. Biol., 42, 1 (1969).
5. S. Sarkar, F. A. Streter, and J. Gergely, Proc. Natl. Acad. Sci. U. S. A., 68, 946 (1971).
6. S. Lowey and D. Risby, Nature, 234, 81 (1971).
7. S. Sarkar, Cold Spring Harbor Symp. Quant. Biol., 37, 14 (1972).
8. F. A. Streter, S. Sarkar, and J. Gergely, Nature (New Biol.), 239, 124 (1972).
9. J. Pollard and R. R. Weihring, Critical Rev. Biochem., 2, 1 (1974).
10. R. V. Rice and A. C. Brady, Cold Spring Harbor Symp. Quant. Biol., 37, 429 (1972).

11. N. A. Rubenstein, J. C. H. Chi, and H. Holtzer, Biochem. Biophys. Res. Commun., 57, 438 (1974).
12. S. M. Heywood and A. Rich, Proc. Natl. Acad. Sci. U. S. A., 59, 590 (1968).
13. S. Sarkar and P. H. Cooke, Biochem. Biophys. Res. Commun., 41, 918 (1970).
14. R. B. Low, J. N. Vournakis, and A. Rich, Biochemistry, 10, 1813 (1971).
15. S. M. Heywood and M. Nwagwu, Biochemistry, 8, 3839 (1969).
16. A. W. Rourke and S. M. Heywood, Biochemistry, 11, 2061 (1972).
17. W. C. Thompson, E. A. Buzash, and S. M. Heywood, Biochemistry, 12, 4559 (1973).
18. G. Brawerman, Ann. Rev. Biochem., 43, 621 (1974).
19. H. Bloemendal, in The Mechanism of Protein Synthesis and its Regulation, L. Bosch, ed., North-Holland, Amsterdam, 1972, p. 487.
20. S. Sarkar, S. P. Mukherjee, A. Sutton, H. Mondal, and V. Chen, Prep. Biochem., 3, 583 (1973).
21. H. Mondal, A. Sutton, V. Chen, and S. Sarkar, Biochem. Biophys. Res. Commun., 56, 988 (1974).
22. G. E. Morris, E. A. Buzash, A. W. Rourke, K. Tepperman, W. C. Thompson, and S. M. Heywood, Cold Spring Harbor Symp. Quant. Biol., 37, 535 (1972).

23. A. Pryzbala and R. C. Strohman, Proc. Natl. Acad. Sci. U. S. A., 71, 662 (1974).

24. H. Aviv and P. Leder, Proc. Natl. Acad. Sci. U. S. A., 69, 1408 (1972).

25. M. Adesnik, M. Salditt, W. Thomas, and J. E. Darnell, J. Mol. Biol., 71, 21 (1972).

26. G. Brawerman, J. Mendecki, and S. Y. Lee, Biochemistry, 11, 637 (1972).

27. P. A. Kitos, G. Saxon, and H. Amos, Biochem. Biophys. Res. Commun., 37, 204 (1969).

28. W. E. Loening, Biochem. J., 113, 131 (1969).

29. R. E. Lockard and J. B. Lingrel, Biochem. Biophys. Res. Commun., 47, 1426 (1972).

30. F. A. Streter, S. Holtzer, J. Gergely, and H. Holtzer, J. Cell. Biol., 55, 586 (1972).

31. A. G. Weeds and S. Lowery, J. Mol. Biol., 61, 701 (1971).

32. M. A. Raftery and R. D. Cole, J. Biol. Chem., 241, 3457 (1966).

33. M. E. Elzinga, Biochemistry, 9, 1365 (1970).

34. S. M. Heywood, D. S. Kennedy, and A. J. Bester, Proc. Natl. Acad. Sci. U. S. A., 71, 2428 (1974).

35. R. Haselkorn and L. B. Rothman-Denes, Ann. Rev. Biochem., 42, 397 (1973).

36. D. T. Wigle, Eur. J. Biochem., 35, 11 (1973).

37. V. Nudel, B. Lebleu, and M. Revel, Proc. Natl. Acad. Sci. U. S. A., 70, 2139 (1973).

38. J. B. Gourdon, C. D. Lane, H. R. Woodland, and G. Marbaix, Nature, 233, 177 (1971).

39. A. J. M. Berns, M. Van Kraaikamp, H. Bloemendal, and C. D. Lane, Proc. Natl. Acad. Sci. U. S. A., 69, 1606 (1972).

40. S. Sarkar, in Exploratory Concepts in Muscle, A. T. Milhorat, ed. Vol. II, Excerpta Medica, Amsterdam, 1974, p. 172.

41. S. Staynov, J. C. Pinder, and W. B. Gratzer, Nature (New Biol.), 235, 108 (1972).

42. U. L. Raj Bhandari, J. Biol., Chem., 243, 556 (1968).

43. J. Nakazato, D. W. Kopp, and M. Edmonds, J. Biol. Chem., 248, 1472 (1973).

44. U. E. Loening, J. Mol. Biol., 38, 355 (1968).

45. M. E. Buckingham, D. Caput, A. Cohen, R. G. Whalen, and F. Gros, Proc. Natl. Acad. Sci. U. S. A., 71, 1466 (1974).

46. S. J. Kaufman and K. W. Gross, Biochim. Biophys. Acta, 253, 133 (1974).

47. S. Petrovic, A. Novakovic, and J. Petrovic, Biochim. Biophys. Acta, 254, 493 (1971).

48. G. Blobel and D. Sabatini, Proc. Natl. Acad. Sci. U. S. A., 68, 390 (1971).

49. H. Holtzer, I. W. Sanger, H. Ishikawa, and K. Strachs, Cold Spring Harbor Symp. Quant. Biol., 37, 549 (1972).

50. R. DeWachter and W. Fiers, J. Mol. Biol., 30, 507 (1967).

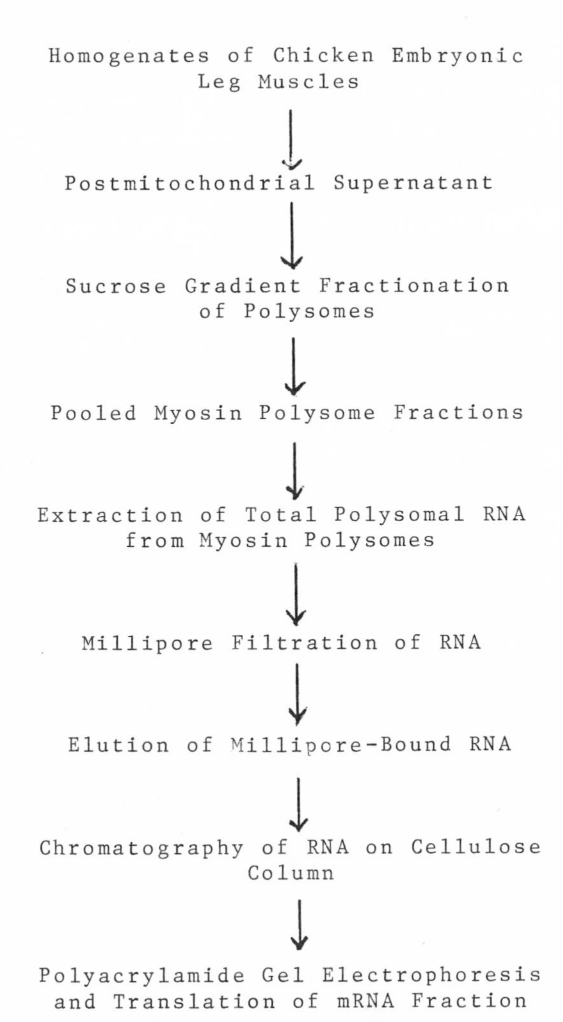


FIG. 1. Scheme for the isolation of myosin polysomes and heavy chain mRNA.

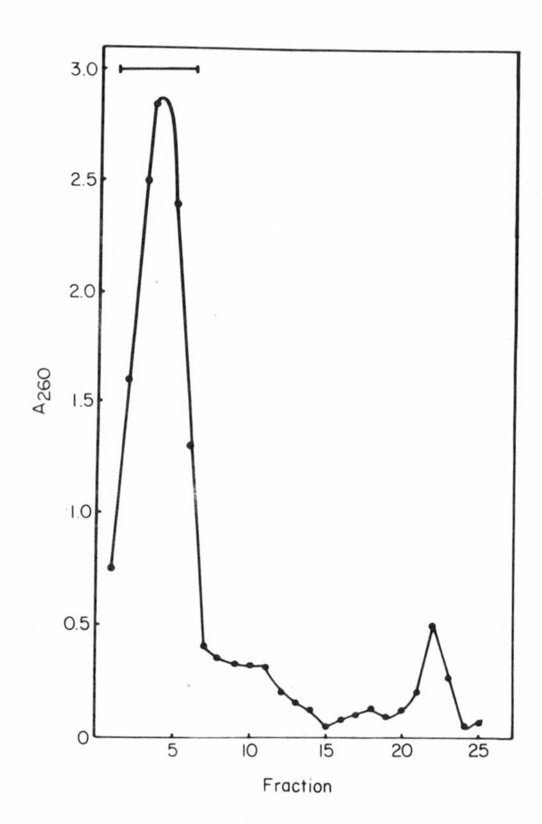


FIG. 2. Sucrose density gradient sedimentation pattern of myosin polysomes. For details, see text. The bottom six fraction (1-6), as indicated by the bar, were pooled and processed for the isolation of heavy chain mRNA. (Reprinted from Ref. 20, p. 587, by courtesy of Marcel Dekker, Inc.)

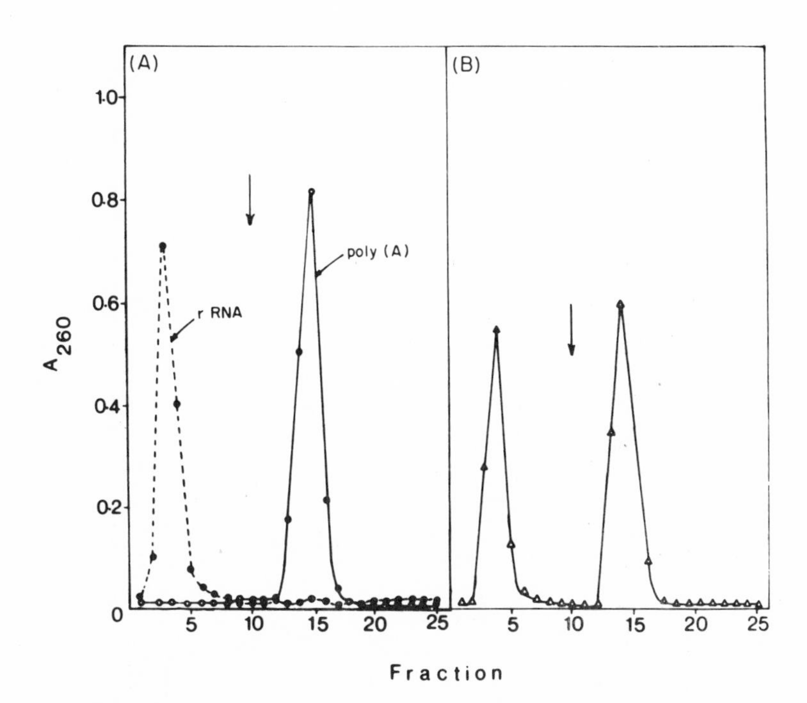


FIG. 3. Chromatography of Millipore-bound RNA fraction obtained from myosin polysomes (Figure 2) on a column of cellulose. For details, see text. (A) Calibration of column with rRNA and poly(A). (B) Chromatography of Millipore-bound RNA. Arrows indicate the position when the eluant was switched to a buffer of low ionic strength; 1-ml fractions were collected.

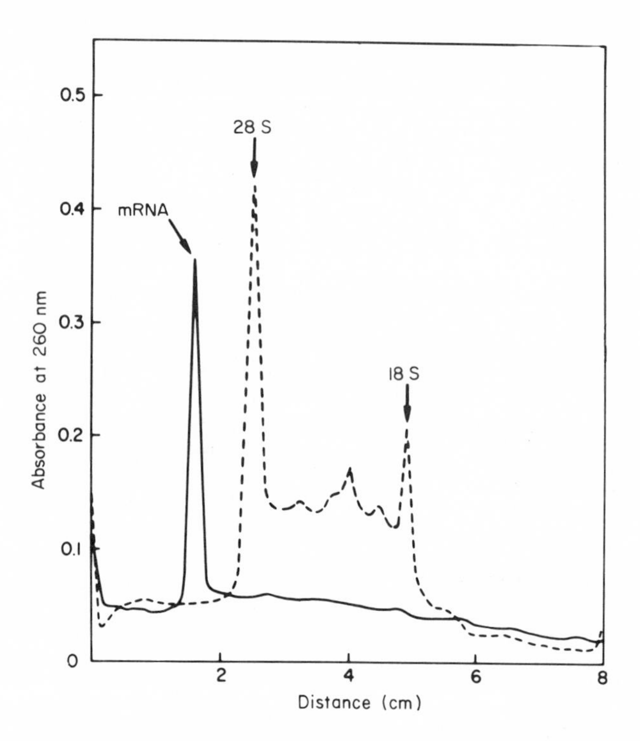


FIG. 4. Densitometric scans of polyacrylamide gel runs of RNA samples eluted from cellulose column (Figure 3). For details see text; 31 μg of RNA from unbound fraction (- - -); 25 μg of RNA bound to and eluted from the column (——).

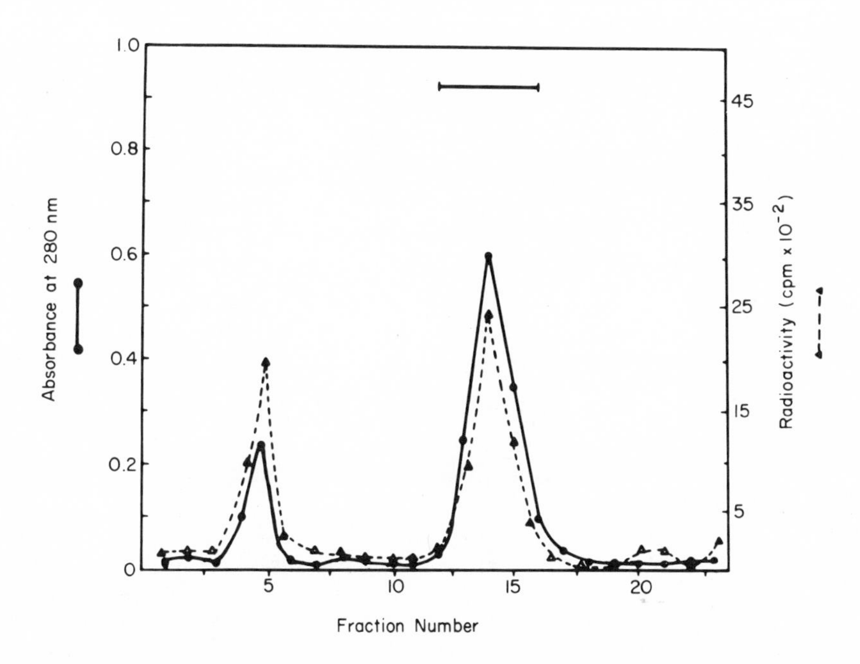


FIG. 5. DEAE-cellulose chromatography of products of in vitro translation of heavy chain mRNA. For details, see text. Fractions indicated by the bar were pooled and concentrated by Aquacide for SDS gel electrophoresis.

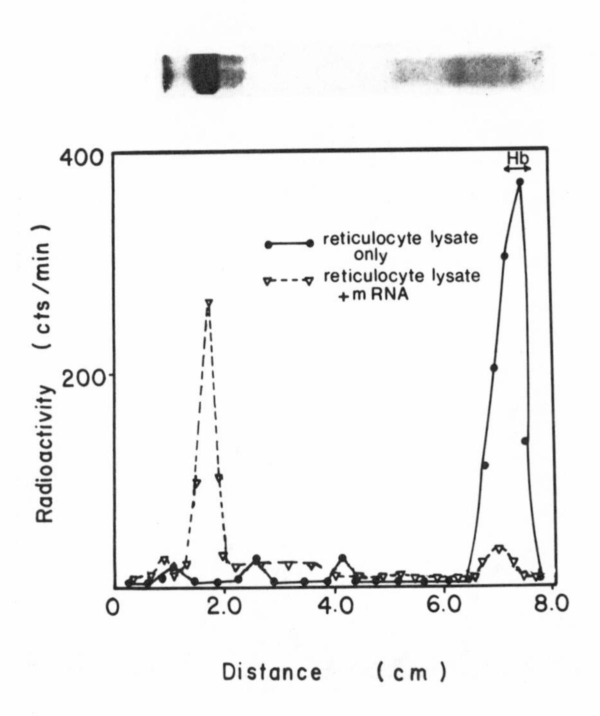


FIG. 6. Electrophoresis of products of in vitro translation of heavy chain mRNA. For details, see text. The gels were stained with Coomassie Brilliant Blue, sliced and counted [13, 20]. The electrophoretogram at the top is obtained by running a sample of 200 μg of purified chicken leg myosin [13] as a marker simultaneously in a separate gel.

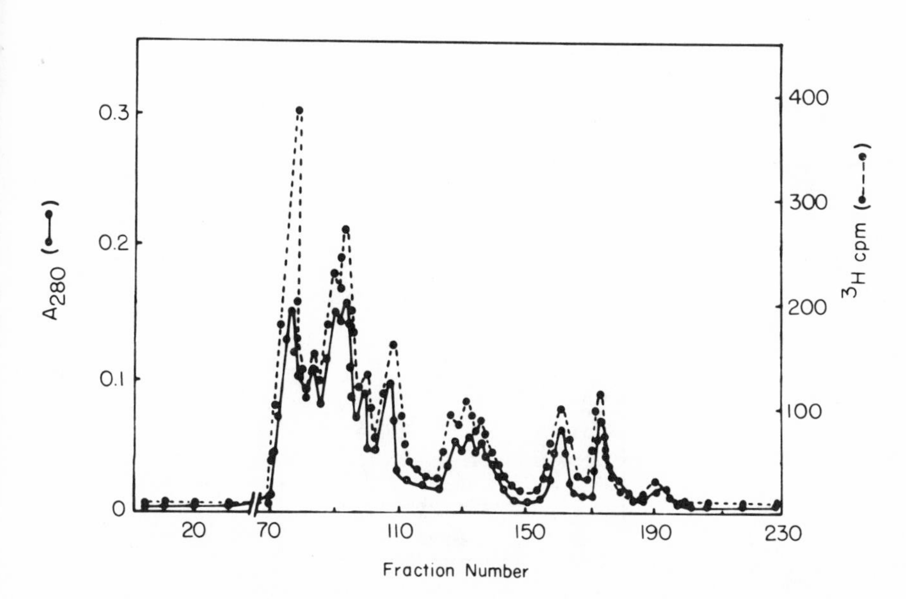


FIG. 7. Analysis of CNBr-peptides of radioactive products of in vitro translation of heavy chain mRNA. For details, see text. After the absorbance at 280 nm was read, the fractions (1 ml each), were dried in scintillation vials and counted using toluene-based counting fluid.

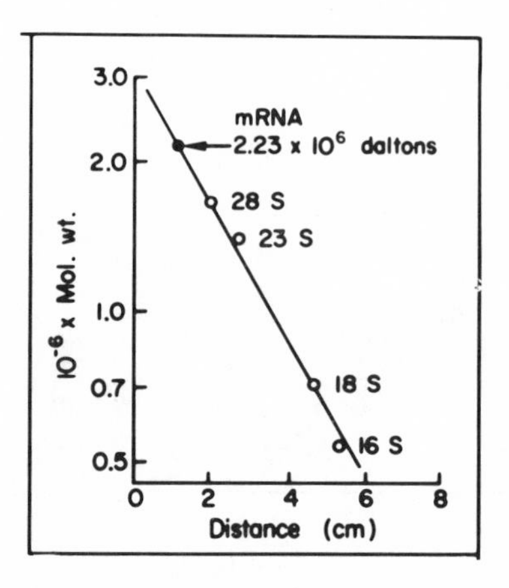


FIG. 8. Determination of molecular weight of heavy chain mRNA. For details see text.

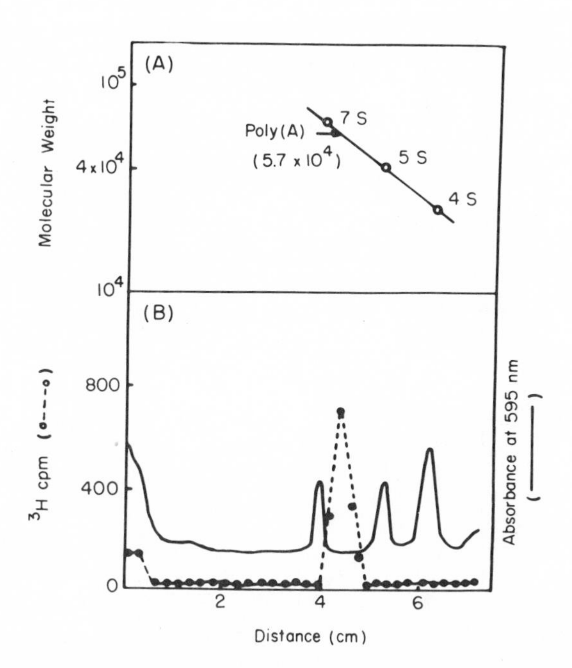


FIG. 9. Estimation of the size of the poly(A) fragment of heavy chain mRNA. For details, see text. Samples of 5S, 4S, and 7S (obtained by heating chicken leg muscle 28S rRNA at 50° C for 20 minutes) RNA and of the tritiated poly(A) fragment were run on 6% gels. After staining the gels with toluidine blue to locate the position of the markers, the gels were sliced and counted [13]. (B) Distribution of radioactivity and UV absorbance. (A) Plot of log molecular weight vs. distance migrated in the gel. (Reprinted from Ref. 21, p. 993, by courtesy of Academic Press, Inc.)

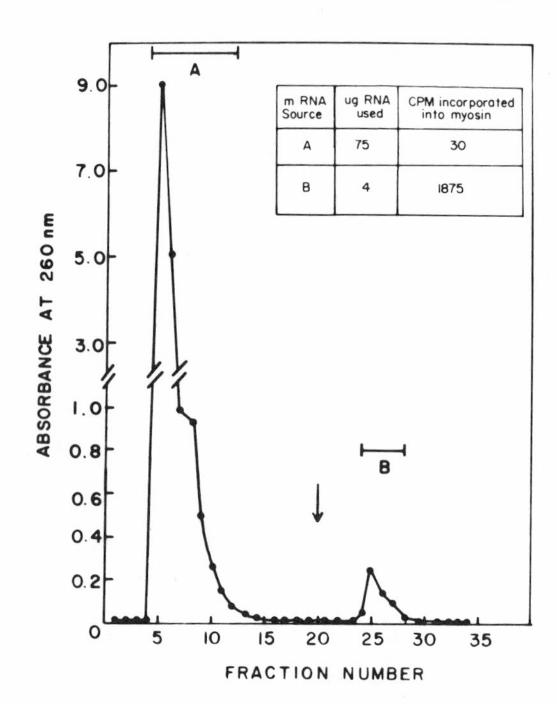


| m RNA Source | ug RNA used | CPM incorporated into myosin |
|---|---|---|
| A | 75 | 30 |
| B | 4 | 1875 |

FIG. 10. Chromatography of polysomal RNA of myosin polysomes on a column of oligo(dT)-cellulose. For details, see text. Twenty-four $A_{260}$ units of polysomal RNA obtained from myosin polysomes and dissolved in 0.01M Tris-HCl (pH 7.6), 0.5M KCl were applied to a column, 1.0 x 0.5 cm, of oligo(dT)-cellulose (type $T_3$, Collaborative Research, Inc.) that had been equilibrated with the same buffer. The column was washed with 20 ml of the same buffer, and then the elution buffer, 10 mM Tris-HCl (pH 7.6), was applied (arrow). The peak fractions, bars A and B, were pooled and precipitated with ethanol (see text). Insert: mRNA activity for myosin heavy chain synthesis.

# CHAPTER 7

# PURIFICATION AND CHARACTERIZATION OF OVALBUMIN mRNA IN THE CHICK OVIDUCT

J. M. Rosen, S. L. C. Woo, A. R. Means
and B. W. O'Malley
Department of Cell Biology
Baylor College of Medicine
Houston, Texas

CONTENTS

## I. INTRODUCTION

Many of the recent advances in the area of steroid hormone action have resulted from the availability of new techniques to study the expression of specific genes during hormone-mediated growth and differentiation directly. Two methodologic advances were of particular importance: First, the development of heterologous cell-free translation systems, in which the biologic activity of an exogenous messenger RNA (mRNA) could be determined [1, 2]; second, the observation that the majority of mammalian mRNAs contain a long sequence of polyadenylic acid residues at their 3'-hydroxyl end [3] permitted the subsequent isolation of these poly(A)-containing mRNAs from the bulk of the stable cellular RNAs by affinity chromatography [4] or adsorption to nitrocellulose filters [5].

Chick oviduct has proved to be an especially useful model system for the study of hormonal regulation of specific gene function [6] . Administration of estrogen to a newborn chick results in the differentiation of three distinct epithelial cell types from the homogeneous population of primitive mucosal cells

[7-9]. Two of these new cell types, the tubular gland cells and the goblet cells, synthesize the cell-specific proteins, ovalbumin and avidin, respectively. Estrogen-inducible ovalbumin constitutes 50-60% of the total protein in a fully differentiated oviduct. Thus, the hen oviduct would be expected to contain high levels of ovalbumin mRNA and its purification should be greatly facilitated. In addition, the other principal egg-white proteins, conalbumin, lysozyme, and ovomucoid, comprise a large percentage of the remaining protein synthesized in the fully differentiated oviduct and their molecular weights differ considerably from the molecular weight of ovalbumin. Accordingly, the sedimentation properties of each of their mRNAs on sucrose gradients are significantly different from those of ovalbumin mRNA [10]. Ovalbumin mRNA routinely sediments in the 16 to 18S region of the gradients [11, 12]. Furthermore, ovalbumin mRNA has also been shown to contain a poly(A) sequence, presumably at its 3' end [11]. A purification scheme based on a combination of precise sizing techniques and affinity chromatography of poly(A)-containing mRNA should, therefore, have a reasonable probability of generating highly purified ovalbumin mRNA.

In this chapter, we will summarize the recent methods that have been utilized in our laboratory and others to purify ovalbumin mRNA, with special emphasis on those methods that are applicable on a preparative scale to the isolation of milligram quantities of the purified mRNA.The physical, chemical and biologic methods used to characterize purified ovalbumin mRNA are also discussed. Many of the criteria used to assess the purity of ovalbumin mRNA should be generally applicable to determining the purity of any isolated mRNA. The effects of estrogen on the induction of ovalbumin mRNA will not be discussed, since the results of these studies have been recently reviewed [13], and methodology similar to that used in our studies will be described elsewhere in this volume.

## II. METHODS OF RNA EXTRACTION

### A. Whole-Cell Extracts

Whole-cell RNA extracts have proved to be an easily obtainable and highly suitable starting material for the further purification of ovalbumin mRNA. Homogenization of frozen tissue directly in phenol and sodium dodecyl sulfate (SDS) helps minimize RNase activity and the consequent loss of mRNA activity. This is a critical problem in subcellular fractionation procedures for polysome isolation (see Sec. II. B). In order to minimize exogenous RNase contamination during the isolation and subsequent purification of ovalbumin mRNA, all glassware is rinsed in 1N NaOH, all solutions are filtered through 0.45 μm Millipore filters (HAWP), and disposable gloves are worn at all times.

The magnum portion of hen oviducts are removed, rinsed in cold saline, and quickly frozen on dry ice. Frozen tissue is stored in liquid nitrogen to prevent the loss of mRNA activity that may occur during partial defrosting in conventional refrigerator freezers. The tissue is then broken into small pieces, and 5 g portions are homogenized at room temperature for one min. in a small Waring blendor (200 ml capacity, model 5019)at maximum speed in five volumes (v/w) of 0.5% SDS, 25 mM $Na_2$ EDTA, 75 mM NaCl (pH 8.0), and five volumes of buffer-saturated, redistilled phenol (pH 8.0). The resulting emulsion is chilled for 30 minutes in ice and centrifuged in 150-ml glass bottles at 6500 x g for 30 minutes. The aqueous upper phase and protein interphase are removed and reextracted with an equal volume of buffer-saturated phenol (pH 8.0) for five min. at room temperature. After centrifugation, the aqueous phase is carefully removed, made 0.2M in NaCl, and overlayed with an equal volume of cold 95% EtOH. Most of the DNA is then removed by spooling onto a glass rod, and the remaining RNA is precipitated by the addition of a second volume of cold ethanol and stored at $-20^{o}$ C overnight. This total RNA extract is reprecipitated from 0.1M NaAc buffer (pH 5.0) several times, rinsed with 100% EtOH, and dissolved in cold distilled water. The total extract may still contain 5 to 10% DNA as measured by the Burton diphenylamine assay. Between 80 and 100 grams of tissue can be processed at one time and routinely yield between 400 and 500 mg of total RNA extract.

In this procedure, the size and shape of the blendor, and the tissue used as the source of RNA, will both markedly affect the amount of contaminating DNA in the RNA preparation and the yield of RNA obtained. Excessive shearing may lead to the extraction of large quantities of DNA, which is normally trapped in the large protein interphase during extraction at room temperature, even at pH 8.0. Moreover, the DNA may be sheared into small pieces and, therefore, may not be removed by spooling. On the other hand, if homogenization is insufficient, large amounts of RNA may be trapped in the protein interphase. Thus , the recovery of RNA and the amount of DNA contamination should be carefully monitored for each blendor and tissue used. For example, this procedure has been readily applied to the isolation of total lactating mammary gland RNA containing less than 10% DNA contamination even without the removal of DNA by spooling. This total RNA extract proved to be suitable for the isolation and purification of casein mRNA [14]. In contrast, under similar conditions rat testis and rat uterine RNA preparations were heavily contaminated with DNA, i.e., 30 to 40%. Removal of contaminating DNA from mRNA preparations can, however, be accomplished by several methods including (dT)-cellulose and Sepharose 4B chromatography (see below).

The use of an acidic pH during phenol extraction will also minimize the extraction of DNA that normally occurs at pH 8.0. Ovalbumin mRNA may be extracted by homogenization of the pulverized frozen tissue in five volumes (w/v) of 10 mM NaAc (pH 5.2), 10 mM $Na_2$ EDTA, 0.5% SDS containing five volumes of buffer-saturated phenol in a small Waring blendor, as described. Prior to the second extraction of the aqueous phase and interphase, NaCl is added to a final concentration of 0.5M to help dissociate any RNA trapped in the nucleoprotein interphase. Comparable yields of both total RNA and ovalbumin mRNA activity are obtained by RNA extraction at either acidic or alkaline pH. Furthermore, the removal of additional DNA from the total extract will prevent the inhibition of translation activity that is usually observed at high RNA concentrations. This permits a more accurate quantitation of the ovalbumin mRNA activity present in the whole-cell RNA extract. The removal of DNA may also improve the efficiency of peptide chain completion or release. For example, the ratio of ovalbumin to total protein synthesized in the whole wheat germ S-30 (see Sec. IV. A) observed with the pH 5.2 RNA is comparable to that found in the postribosomal supernatant with the pH 8.0 RNA extract.

Since the recovery of poly(A)-containing mRNA during phenol extraction is influenced by such factors as pH, ionic strength, and protein concentration [5], it is important to monitor mRNA recovery when these procedures are applied to other tissues or sources of RNA, i.e., nuclei or polysomes. For example, considerable loss of poly(A)-containing mRNA has been reported to occur when polysomes are extracted with phenol alone at neutral pH [15]. This loss could be overcome by the addition of chloroform during the phenol extraction procedure or by Pronase treatment before extraction. Extraction of RNA with a mixture of chloroform and phenol also improves the separation and reduces the amount of the aqueous phase retained in the organic phase. In tissues with high lipid contents, such as the lactating mammary gland, chloroform addition is also advantageous. Homogenization is initially performed with only the phenol-SDS buffer solution, then chloroform is added during the second extraction of the aqueous phase and interphase.

Finally, the extraction of ovalbumin mRNA from whole tissue using only detergents has been reported [16]. The omission of phenol has the advantage of preventing mRNA aggregation that may result from phenol extraction [17]. This method relies on deproteinization with 1% SDS, followed by sucrose gradient ultracentrifugation to separate RNA from the majority of protein and heparin, added as a ribonuclease inhibitor. The details of this procedure have been published [16]. Its principal disadvantages are the incomplete removal of protein from RNA with SDS alone and the necessity of using ultracentrifugation, which markedly limits the amount of tissue or polysomes that can be processed at a given time. Since methods are now available to prevent ovalbumin mRNA aggregation [18, 19], (Sec. III. A), the omission of phenol during RNA extraction appears to be of limited utility.

### B. Polysomal RNA

Ovalbumin mRNA has been successfully isolated by the procedures outlined from polysomes, as well as from whole tissue. Total polysomes obtained by either sucrose gradient ultracentrifugation [19, 20] or $Mg^{2+}$ precipitation [21], and polysomes containing nascent ovalbumin chains isolated by selective immunoprecipitation [22, 23] have all been used as starting materials for the subsequent isolation of ovalbumin mRNA.

In each of these procedures the polysome-containing, postmitochondrial supernatants are initially prepared in essentially the same manner [24]: The epithelial cell layer removed by scraping from fresh oviduct or obtained by passage of the oviduct through a Harvard Tissue press is homogenized in 7 volumes (w/v) of polysome buffer (25 mM Tris-HCl, pH 7.6, at 4° C, 25 mM NaCl, 5 mM $MgCl_2$, 140 mM sucrose, and sodium heparin (0.5-1 mg/ml), using a Dounce homogenizer. Six strokes are made with the A pestle, then two volumes of 5% sodium deoxycholate and 5% Triton-X-100 are added, and homogenization is continued with three strokes of the A pestle and two strokes of the B pestle. The percentages of these two detergents has been slightly varied in this technique, and, in some cases, a final concentration of 2% Triton-X-100 alone is also used [21]. The homogenate is then centrifuged for five minutes at 27,000 g in a refrigerated centrifuge, and the supernatant is used for polysome isolation.

Total polysomes may be isolated by layering 5-ml aliquots of the supernatant over a discontinuous sucrose gradient containing 2 ml of 2.5M sucrose and 4 ml of 1.0M sucrose, prepared in polysome buffer containing 40 μg/ml of sodium heparin. Ultracentrifugation is then performed in a Beckman 60 Ti rotor for 2.5 hours at 50,000 rpm. Alternatively, the ribosomes in the supernatant may be precipitated, without the use of ultracentrifugation, by the addition of $MgCl_2$ to 100 mM [21]. After a one hour incubation at $0^o$ C, the supernatants are layered over 4-ml pads of 0.2M sucrose in a modified polysome buffer containing 100 mM $MgCl_2$ and centrifuged at 27,000 g for 10 minutes. This method is rapid and well suited for large scale isolation of polysomes and subsequent mRNA isolation.

A priori it would be expected that a selective immunoprecipitation method would be the technique of choice for the isolation of a highly purified mRNA. This method depends on the ability of a highly purified, RNase-free antibody to specifically recognize nascent polypeptide chains present on intact polysomes. In addition, the amount of trapping during immunoprecipitation should be minimal in order to limit the contamination with other mRNAs. Finally, since the specific mRNA will comprise only a small percentage of the total polysomal RNA (less than 1%), the technique should be applicable to large-scale preparations in order to permit the isolation of adequate quantities of the mRNA, free of ribosomal RNA, transfer RNA, and any nuclear RNAs that may be present in the polysome preparations.

Three different methods of immunoprecipitation have been successfully used by Schimke and his collaborators [22, 23, 25] for the isolation of ovalbumin mRNA-containing polysomes: (a) In their initial method, the direct immunoprecipitation of ovalbumin and an ovalbumin serum [25]. (b) This technique was improved by the immunoadsorption of ovalbumin-synthesizing polysomes to a cross-linked ovalbumin matrix [22]. (c) The most recent modification of this procedure uses an indirect immunoprecipitation technique, in which the soluble antibody-nascent chain-polysome complex formed by incubating antibody with polysomes is then precipitated by reacting with an anti-antibody [23]. The details of each of these procedures are described in the references cited. Both the direct immunoprecipitation and the immunoadsorption techniques have the disadvantage of requiring large amounts of purified antigen and RNase-free antibody. In addition, they result in both low recoveries of polysomes (maximum of 30%) and relatively high levels of nonspecific trapping (2.5 to 10%). These limitations have been partially overcome by the use of indirect immunoprecipitation. Nonspecific trapping has been reduced to approximately 0.4%, and the yield of polysomes is increased to 60 to 70%. Considerably smaller amounts of highly purified antigen and antibody are also required.

The preparation of a high-affinity, ribonuclease-free antiserum is an essential prerequisite for all immunoprecipitation procedures. This requires the purification of a specific antibody by affinity chromatography, followed by one or two passages of the purified antibodies over sterile columns of CM-cellulose and DEAE-cellulose [23]. Furthermore, isolation of polysomes containing undegraded mRNA is also obviously critical. In some tissues, this necessitates the use of high levels of RNase inhibitors during subcellular fractionation procedures. Each of the oviduct polysome isolation procedures that have been described requires high levels of heparin, added as an RNase inhibitor, to prevent ovalbumin mRNA degradation. Heparin is usually added at concentrations of 0.5 to 1 mg/ml to all homogenization buffers. But, heparin must then be removed before assaying mRNA translation activity because of its significant inhibitory effect [20]. It may be removed by washing the RNA precipitate obtained after phenol or SDS extraction and ethanol precipitation with 3M NaAc (pH 6) [20, 21]. This procedure removes heparin, low molecular weight RNA, and DNA fragments. After the 3M NaAc washing procedure, the RNA is reprecipitated from 0.1M NaAc buffer as described (Sec. II. A) to remove excess salt that may be trapped in the pellet.

With the availability of improved polysome isolation procedures and immunoprecipitation techniques, these methods are now potentially applicable to the isolation of larger amounts of purified ovalbumin mRNA. Previously, they have only generated microgram quantities of a partially-purified mRNA. These techniques, however, should be especially suitable for the isolation of small quantities of those mRNAs that are not unique in size and that code for proteins representing a small percentage of total cellular protein synthesis. As previously mentioned, this is not the case for ovalbumin and its mRNA. Thus, the use of immunoprecipitation techniques and polysomes as the source of ovalbumin mRNA is of limited advantage. Milligram quantities of purified ovalbumin mRNA, free of the stable cellular RNAs (rRNA, tRNA) and other potential nuclear RNA and mRNA contaminants, can be easily isolated from a whole-cell RNA extract by a combination of precise sizing techniques and affinity chromatography [14]. For studies in which the poly(A)-containing ovalbumin mRNA is used as a template for the oligo(dT)-primed, RNA-directed DNA polymerase reaction, small amounts of ribosomal RNA contamination are not critical. In these studies polysomal RNA has proved to be a useful starting material for the partial purification of ovalbumin mRNA. The resulting complementary DNA sequence is a specific and sensitive hybridization probe and a powerful new tool with which the regulation of ovalbumin mRNA synthesis, processing, and turnover has been studied [26, 27, 28]. However, large amounts of uncontaminated ovalbumin mRNA are essential for future gene isolation experiments and for accurate sequence and secondary structure determinations. In these experiments even slight contamination with ribosomal RNA, which is transcribed from repetitive gene sequences, may lead to serious artifacts. Two such examples are early hybridization experiments with hemoglobin and silk fibroin mRNA preparations which were contaminated with rRNA, and which markedly overestimated the frequencies of the hemoglobin and fibroin genes [29, 30].

## III. RNA FRACTIONATION

### A. Sizing Methods

The formation of stable aggregates of ovalbumin mRNA presents a serious problem in both molecular weight estimation and in fractionation procedures designed to yield a discrete band of mRNA activity [19, 31]. As previously mentioned, mRNA aggregation may result from phenol extraction [17] and can be prevented by the use of partially or fully denaturing conditions during fractionation procedures. A similar phenomenon has been observed for immunoglobin light chain mRNA [32] and may, in that case, be advantageous for the further purification of light chain mRNA. However, as a rule, it would appear that mRNA purification procedures, and especially the analytic methods used to assess mRNA purity, should incorporate denaturing conditions. Only then can the separation of mRNAs into discrete molecular species and the identification of any higher molecular weight precursors be accomplished. Such conditions were found to be necessary during the purification of ovalbumin mRNA.

Sucrose Gradient Centrifugation

As previously reported [11, 12, 19], when either oviduct polysomal RNA or a total cell RNA extract are analyzed on linear sucrose gradients, most of the ovalbumin mRNA activity is found in the 16-18S region of the gradients. In some experiments, however, as much as 40 to 50% of the ovalbumin mRNA activity is found as higher molecular weight aggregates [12, 19, 31]. Although zonal ultracentrifugation has proved to be an especially useful technique for the isolation of large amounts of hemoglobin mRNA [32], the problem of mRNA aggregation resulted in poor recoveries of ovalbumin mRNA activity during zonal ultracentrifugation under nondenaturing conditions. Attempts to disaggregate this higher molecular weight ovalbumin mRNA activity by treatment with EDTA or SDS, by use of low salt gradients, or by treatment with dimethylsulfoxide($Me_2SO$) followed by reprecipitation proved unsuccessful [31]. This problem could be avoided, however, by rapid heating of the total RNA extract to $70^{\circ}$ C, followed by room temperature, zonal ultracentrifugation on low salt-containing (<10 mM) sucrose gradients.

Zonal ultracentrifugation is performed using a Beckman Ti-14 rotor for 12 hours at 48,000 rpm at $21^{o}$ C. Approximately 30 mg of total RNA is dissolved in 4 mM $NH_4$acetate buffer (pH 7.0) containing 2 mM $Na_2$ EDTA and 1% SDS. The RNA is rapidly heated in this buffer to $70^{o}$ C for five minutes and immediately sedimented through a linear 10 - 40% (w/w) sucrose gradient in 0.1% SDS, 2 mM $Na_2$ EDTA, and 4 mM $NH_4$acetate buffer (pH 7.0). The sample is assayed at a rate of 20 ml/min. and the absorbance is monitored with an ISCO UV-2 monitor at 254 nm. Under these conditions the ovalbumin mRNA activity sediments as a broad band between 16 and 18S and is clearly separable from 4S RNA and from 28S RNA. Other functional mRNAs present in the 9 to 15S and 18 to 28S areas of the gradient are also separated from ovalbumin mRNA as shown by the determination of total mRNA activity in the wheat germ assay. (See below and Chap. 1.) No ovalbumin mRNA activity is found in fractions sedimenting faster than 18S rRNA, indicating the lack of mRNA aggregates.

Ovalbumin mRNA aggregation may also be prevented during sucrose gradient centrifugation by the inclusion of such denaturing solvents as formamide or $Me_2SO$. Formamide-containing sucrose gradients have been an especially useful analytical tool to analyze oviduct nuclear RNA [27] and purified ovalbumin mRNA preparations [18, 34]. Analysis of RNA on 5 - 25% (w/w) linear sucrose gradients containing 70% formamide (Fisher F-95, $A_{270}$ <1.0) and 3 mM $Na_2$ EDTA, 3 mM Tris-HCl (pH 7.4) is performed essentially as described by Suzuki et al. [30]. The RNA samples dissolved in the buffered formamide are incubated at 37$^o$ C for 15 minutes, layered on the gradients and centrifuged for 20 hours at 35,000 rpm in a Beckman SW40 rotor at 25$^o$ C. The RNA may be recovered from the gradient fractions by ethanol precipitation; the pellets are rinsed with 95% EtOH to remove any remaining formamide. When partially purified ovalbumin mRNA is analyzed on formamide-containing sucrose gradients, a sharp peak of ovalabumin mRNA activity is found to sediment at 16S. This coincides with the majority of total mRNA activity, although some additional mRNA activity is also observed in the 6 to 15S region of the gradients. But, when purified ovalbumin mRNA is studied under these same conditions, a single sharp RNA peak sedimenting at 16S is observed, which is essentially superimposable with both the peaks of total mRNA activity and ovalbumin mRNA activity, as determined in the wheat germ assay.

Sepharose 4B Chromatography

Although the preparative fractionation of oviduct RNA on linear sucrose gradients results in a substantial enrichment in ovalbumin mRNA activity, this method is unable to totally remove the sheared DNA present in the total nucleic acid extract. This DNA is observed as a broad band sedimenting between 14 and 28S. In addition, some of the DNA is adsorbed to Millipore filters under the conditions used to bind poly(A)-containing mRNA (see Sec. III. B). The DNA may be removed by chromatography of the total RNA extract on Sepharose 4B in 0.1M NaAc buffer (pH 5.0), containing 1 mM $Na_2$ EDTA. Approximately 100 mg of the total extract, at a concentration of 5 mg/ml, is applied to a 5x 100 cm column and eluted at a flow rate of 30 ml/hour. When 15 mg of the filtered or (dT)-bound RNA is fractionated (concentration, 1 - 2 mg/ml), a 2.6x 100 cm column is used and the flow rate is reduced to 8 ml/hr. Chromatography is usually performed by a reverse flow method to prevent undue gel packing.

Under these conditions, the DNA is excluded from the column, and all the detectable ovalbumin mRNA activity appears in a peak slightly preceding the 18S rRNA peak. As expected, a peak of 4S RNA is eluted after the 18S rRNA peak. But, quite unexpectedly, the elution of 28S rRNA is considerably retarded when the chromatography is performed in the 100 mM NaAc buffer. This allows for the removal of most of the contaminating 28S RNA from ovalbumin mRNA. The result is somewhat fortuitous, since 28S RNA is a major contaminant present after the adsorption of ovalbumin mRNA to nitrocellulose filters [18] or chromotography on (dT)-cellulose [19]. Furthermore, the ovalbumin mRNA is also partially separated from other smaller molecular weight mRNAs that are eluted with the trailing edge of the 18S rRNA peak. Therefore, when RNA fractions obtained from the front side of the mRNA activity peak are analyzed in the wheat germ assay, they synthesize predominantly ovalbumin, as demonstrated by the comigration of the radioactive peptides on SDS gels with a specifically immunoprecipitated ovalbumin standard. Moreover, a progressive increase in the proportion of smaller molecular weight peptides and a corresponding decrease in radioactivity in the region of ovalbumin is observed when RNA fractions from the peak and trailing side of the mRNA activity peak are assayed. Thus, different-sized classes of mRNA can be separated by chromatography on Sepharose 4B and a selective enrichment of ovalbumin mRNA (approximately six to eight-fold compared to the total RNA extract) can be obtained by pooling the appropriate fractions between the DNA and 18S peaks. Sepharose 4B chromatography does not require the sophisticated instrumentation necessary for zonal ultracentrifugation and also permits the removal of most of the contaminating DNA from ovalbumin mRNA. By this procedure, 100 mg of a total RNA extract can be easily and reproducibly fractionated. This technique can also be utilized for the separation of mRNAs that have been previously enriched by adsorption to nitrocellulose filters or (dT)-cellulose chromatography. In our laboratory, Sepharose 4B chromatography is usually used as the second step of ovalbumin mRNA purification, following the poly(A)-adsorption techniques. This allows for the more rapid purification of large amounts of ovalbumin mRNA, since 1 g of total RNA can be processed on large (dT)-cellulose columns more rapidly than by Sepharose 4B chromatography.

Agarose gel electrophoresis

Throughout the purification scheme, RNA samples are characterized by electrophoresis on 1.5% agarose gels containing 6M urea and 25 mM citric acid (pH 3.5). Electrophoresis grade agarose was obtained from BioRad and is dissolved by gentle refluxing in the urea buffer (see below). Gels (8 cm in length) are poured in 5 mm (I.D.) acid-washed tubes, with their bottoms covered with dialysis tubing and parafilm. The agarose forms a gel in approximately 30 minutes at room temperature, and the gels are then held at $4^{o}$ C until they become opaque. Since no polymerizing agents are used, electrophoresis to remove potentially harmful reagents is probably unnecessary but it is usually performed at 2 mA/tube for 15 minutes to allow equilibration with the electrophoresis buffer. The RNA samples are dissolved in a 6M urea, 25 mM sodium citrate layering buffer containing 20% sucrose (RNase-free, Schwarz/Mann) and 0.005% Bromphenol Blue (BPB). The electrophoresis buffer is composed of only 25 mM citric acid (pH 3.5). Electrophoresis is then performed for 7 hr. at 2mA/gel at $2^{o}$ C in a Buchler polyanalyst. Gels are stained for 30 minutes in a 1% methylene blue-15% HAc solution, destained in water and scanned at 600 nm, using a Gilford 2400S spectrophotometer equipped with a Model 2410S linear transport. A slightly uneven background in the agarose gel, rather than discrete RNA bands, accounts for the occasional baseline variability.

Agarose gel electrophoresis in 6M urea (at pH 3.5) has several advantages over polyacrylamide gel electrophoresis. Agarose gels require no polymerizing agents that may affect RNA. They are easily poured, physically firmer than low percentage acrylamide gels, and directly adaptable to preparative gel electrophoresis [34, 35]. In addition, agarose gel electrophoresis permits the fractionation of high molecular weight RNA under denaturing conditions that minimize mRNA aggregation with comparable resolution to that obtained using polyacrylamide gels. Although 6M urea does not appear to be an effective denaturing agent at neutral pH, no mRNA aggregates are found under the acidic conditions used. No significant effect of the protonation of cytosine or adenosine residues at pH 3.5 was observed on the migration of high molecular weight RNA during agarose gel electrophoresis. Finally, small amounts of contaminating high molecular weight DNA do not clog the agarose gels as often happens with polyacrylamide gels of greater than 3%.

When ovalbumin mRNA activity is to be measured in the wheat germ assay, the RNA may also be extracted from unstained gels. Gels are initially scanned at 270 nm and then sliced manually with a razor blade. The slices are individually homogenized in cold 0.1M Na acetate buffer (pH 5.0), using a small Teflon pestle homogenizer. After centrifugation at 4° C for 15 minutes at 20,000 g, the aqueous upper layer is carefully removed from any pelleted agarose and brought to 0.5M $Na^+$. The RNA is then precipitated by the addition of a second volume of 100% EtOH and stored at $-20^o$ C overnight. The resulting RNA is reprecipitated from acetate buffer, rinsed with 95% EtOH and dissolved in distilled water. Although recovery of both mass and mRNA activity from gels by this procedure is low (approximately 5% of the starting RNA), there is adequate activity to identify ovalbumin mRNA in the wheat germ assay. Efficient extraction of microgram quantities of functional mRNA from either polyacrylamide or agarose gels, especially when denaturing conditions are used, has proved to be especially difficult. These difficulties can be overcome by the use of preparative gel elcttrophoresis.

Since agarose gels have extremely low adhesiveness and cannot remain in the gel electrophoresis column without support, we designed an adapter to the Buchler Poly-Prep Electrophoresis Apparatus for support of the gel. The adapter is made of a 6.0 cm tall Lucite column (O.D. 5.0 cm; I.D. 4.2 cm), which fits precisely into the gel electrophoresis column. It is tightly held in the gel column by two Viton O-rings, fitted over two grooves engraved on both ends of the adapter. A piece of Nytex screen is secured at the bottom of the adapter with the aid of a snap ring. The adapter was fitted into the gel column, and the gel is cast inside the adapter. The preparative gel procedure is as follows: A 2% agarose solution is made by gently refluxing 1.0 g of agarose in 50 ml of 25 mM Na citrate (pH 3.5), containing 6M urea, for 10 min. with magnetic stirring. The slightly opaque solution is allowed to cool for 15 minutes in a 62° C water bath, and 40 ml of it is transferred, using a separatory funnel, into the adapter in the gel column. The gel solution in the adapter is slowly cooled to 4° C by regulating the temperature of coolant, which circulates continuously through the column jackets. Overlayering the gel surface with water is avoided. An opaque agarose gel forms in the adapter within one to two hours. The gel-forming insert is then replaced by the glass membrane holder. An elution chamber 1 mm deep is established by retracting the glass membrane holder one-half turn. The lower electrophoresis chamber is then filled with 25 mM Na citrate (pH 3.5), and the upper chamber is filled with 50 mM Na citrate, (pH 3.5). The elution chamber is filled with the elution buffer (0.125M Na citrate, pH 3.5, containing 6M urea and 250 g/liter of sucrose). It is imperative not to raise the elution buffer level higher than that of the upper chamber buffer, otherwise the agarose gel will float in the gel column due to the increased hydraulic pressure. The gel is then subjected to preelectrophoresis at 4° C for 30 min. at a constant current of 35 mA. The RNA (3 - 4 mg) is mixed with three volumes of layering buffer (25 mM sodium citrate, pH 3.5, containing 8M urea, 200 g/liter of sucrose, and 0.5 g/liter of Bromphenol Blue), and the entire solution is applied to the top of the agarose gel using a peristaltic pump at low speed. Electrophoresis is immediately resumed at 35 mA. The voltage is generally 50-80 V throughout electrophoresis. The RNA is eluted from the elution chamber by drawing the elution buffer through the chamber at a rate of 0.5 ml/min, using a peristaltic pump. It is necessary to draw buffer through the elution chamber, because forcing buffer into the chamber will cause the gel to float. Fractions of 8 ml are collected, and the RNA in each fraction is precipitated with alcohol. Precipitates are collected by centrifugation and washed once with 10 ml of 66% cold EtOH containing 0.1M NaCl. The washed pellets are dried and redissolved in 0.1 ml of water. Ovalbumin mRNA activity and total mRNA activity are then assayed in the wheat germ translation system.

### B. Poly(A)-Adsorption Techniques*

Most of the stable cellular RNAs may be removed from poly(A)-containing mRNAs by either affinity chromatography on poly(U)-Sepharose [36] or (dT)-cellulose [4], or by adsorption of the poly(A)-containing mRNAs to nitrocellulose filters [5]. Both Millipore filtration and (dT)-cellulose chromatography have been successfully applied in our laboratory [12, 18, 37] and others [19, 22] for the purification of ovalbumin mRNA. These techniques rely on the ability of the poly(A)-containing mRNA to be selectively retained when applied in a high salt buffer, usually 0.5M KCl-0.01M Tris-HCl (pH 7.6). These conditions favor the formation of a stable (A-T) duplex at room temperature or the adsorption of poly(A) to nitrocellulose. Most of the ribosomal RNAs, tRNA, and DNA are removed by washing with the high salt buffer. The mRNA is then eluted by the removal of KCl, either by elution with 0.01M Tris-HCl (pH 7.6) alone for (dT)-cellulose chromatography, or with 0.01M Tris-HCl (pH 7.6) [18] or pH 9.0 [5, 12] , or with 0.01M EDTA (pH 5.0) [22], each containing 0.5% SDS in the nitrocellulose filter method.

---

*(See Chaps. 1, 3, 6)

It should be noted that neither technique alone can completely remove rRNA contamination when a total cell RNA extract or polysomal RNA extract is used in preparative scale procedures. After a single chromatography or adsorption step an analysis of nitrocellulose- and (dT)-bound RNA by agarose gel electrophoresis under denaturing conditions reveals that both fractions are still contaminated with 28S and 18S rRNA, which still comprises almost 60% of the RNA. Some DNA is still present in the Millipore-bound RNA but is absent after (dT)-cellulose chromatography. The presence of DNA in the Millipore-bound RNA is not unexpected, since denatured DNA is known to bind to nitrocellulose filters under the high salt conditions used for mRNA binding. But tRNA and 5S RNA are completely removed from poly(A)-containing mRNA using either of these techniques. When hen oviduct RNA preparations are purified, a new absorbance peak in the 20 to 21S region of the gel is found with both techniques; this peak is well separated from the 18S rRNA peak. This RNA band contains all the ovalbumin RNA activity. Rechromatography of the partially purified mRNA preparation containing approximately 60% of 18S rRNA results in the additional removal of some of this RNA, but the bound ovalbumin mRNA still contains approximately 30% of 18S rRNA.

Oligo (dT)-cellulose chromatography is more selective than Millipore filtration, allowing a 25-fold vs. a 16-fold increase in ovalbumin mRNA activity. It also results in the increased recovery of total ovalbumin mRNA activity (60 - 80% vs. 30 - 50% for Millipore filtration). In both techniques, it is important to determine the capacity of the batch of nitrocellulose filters or the (dT)-cellulose used in order to allow for maximum recoveries (see below). Although (dT)-cellulose chromatography permits an increased recovery of ovalbumin mRNA activity, it is equally effective in the isolation of other contaminating mRNAs. Thus, Millipore filtration results in a larger increase in the ovalbumin mRNA to total mRNA ratio, presumably due to the partial loss of some other mRNAs during the Millipore procedure. This can be explained by the observation that (dT)-cellulose is able to bind mRNAs containing poly(A) tails of 20 residues or longer, whereas Millipore filtration can only retain mRNAs with poly(A) tails longer than 50 adenosine residues. A heterogeneity in the lengths of poly(A) tails that has been observed for several mRNAs [37] may, therefore, result in a somewhat selective purification of ovalbumin mRNA.

Adsorption of poly(A)-containing mRNA to nitrocellulose filters is performed in our laboratory as follows: Approximately 2 mg of the total RNA extract or 100 - 200 μg of the partially purified ovalbumin mRNA is applied per 25 mm, 0.45-μm nitrocellulose filter (Millipore Corporation, type HAWP) in 0.5M KCl, 1 mM $MgCl_2$, 10 mM Tris-HCl (pH 7.6) buffer at room temperature. The total RNA, at a concentration of 200 μg/ml, or the purified RNA, at a concentration of 10 - 20 μg/ml is slowly passed through the filters at a flow rate of approximately 1 drop/sec. The filters are then each washed with 10 to 20 ml of the high salt buffer. In the preparative scale procedure, mRNA is eluted from 100 filters by shaking the filters with 50 ml of 0.1 Tris-HCl (pH 7.6) containing 0.5% SDS for 30 minutes. The filters are removed, and a second identical elution is again performed for 30 min. The combined eluate is chilled in ice for 30 min. to precipitate most of the SDS. Then KCl is added to 0.25M, and the remaining insoluble potassium dodecyl sulfate is quantitatively removed by centrifugation. The RNA is recovered from the supernatant by ethanol precipitation.

Affinity chromatography using (dT)-cellulose is performed at room temperature by the method of Aviv and Leder [4], with the 0.1M KCl intermediate salt wash omitted; 200 mg of the total RNA extract dissolved in 100 ml of buffer (0.5M KCl, 0.01M Tris-

HCl, pH 7.6) is applied at a flow rate of 0.5 ml/min to 10 g of oligo(dT)-cellulose packed in a 1.6 x 15 cm column. Oligo(dT)-cellulose (6 to 18 dT residue average) can be obtained commercially (Collaborative Research, Inc.) (Type $T_3$) or can be synthesized. The capacity of each batch is quite variable and should be determined for each preparation by measuring the amount of mRNA activity in the unbound fraction and assessing its ability to bind to a reequilibrated column. The mRNA fraction is eluted from the column with 0.01M Tris-HCl (pH 7.6), and routinely comprises between 1.5 and 2.5% of the total RNA applied. The column is stored in 0.02% $NaN_3$ at $4^\circ$ C and stripped with 0.1N NaOH between each run. Under these conditions, no loss of binding capacity is observed over a 6 month period.

The previously described sizing and poly(A) purification techniques may, therefore, be utilized in a sequential manner to prepare large amounts of purified ovalbumin mRNA. The initial purification step is usually either (dT)-cellulose chromatography or the adsorption of mRNA to nitrocellulose filters, because of the large amount of RNA that can be rapidly processed and the removal of most of the stable cellular rRNAs and tRNA. This procedure is followed by Sepharose 4B chromatography. The resulting peak of ovalbumin mRNA is then further purified by an additional poly(A) enrichment procedure. A precise sizing technique, i.e., preparative agarose gel electrophoresis, is finally used to yield a single band of RNA. In the following section we will discuss the procedures that are used to assess the purity of the ovalbumin mRNA.

## IV. ASSESSMENT OF PURITY

### A. Biological Methods--Wheat Germ Translation Assay*

Although in most of our previous studies [11, 12] , ovalbumin mRNA activity was measured in a modified reticulocyte lysate system [1], the wheat germ protein synthesis system [38] has several advantages for the assessment of ovalbumin mRNA purity. The wheat germ cell-free system is characterized by an extremely low level of endogenous protein synthesis [39], in contrast to the high levels of endogenous hemoglobin synthesis in the reticulocyte lysate and may therefore be used to assay total mRNA activity in addition to a specific mRNA activity. A comparison of the incorporation of radioactivity into a spe-

*(See also Chap. 1)

cific antibody-precipitable protein with the total incorporation into protein can be used as an assessment of the biologic purity of a given mRNA. In addition, the radioactive polypeptides synthesized in response to an exogenous mRNA can be directly analyzed without the use of immunoprecipitation, by either SDS gel electrophoresis of the released polypeptide chains [40] or by two-dimensional paper electrophoresis of partial typtic digests of the radioactive products [41].

Preparation of the wheat germ 30,000 g supernatant fraction (S-30) is performed by a modification of the method of Roberts and Paterson [38]. Wheat germ may be obtained from General Mills, Inc., Minneapolis, Minn. Our present lot was shipped from Vallejo, Calif. Upon arrival it was immediately stored at $4^{o}$ C in a vacuum desiccator. The S-30 fraction is prepared by homogenization of 20 g of wheat germ at $4^{o}$ C in 50 ml of buffer A (20 mM Hepes, titrated to pH 7.6 with KOH, 0.10M KCl, 1 mM Mg $acetate_2$, 2 mM $CaCl_2$ and 6 mM 2-mercaptoethanol) using a Polytron homogenizer (Brinkmann) set at "5" for one minute. An additional 50 ml of buffer A is added and the homogenization is continued for another 15 secs., with precautions taken to prevent frothing. The homogenate is centrifuged at 30,000 g for 10 minutes and the supernatant is applied to a Sephadex G-25 column (5.0 x 50 cm) preequilibrated with buffer B (20 mM Hepes, pH 7.6, 0.12M KCl, 5 mM Mg $acetate_2$, 6.0 mM 2-mercaptoethanol). The most turbid fractions eluting at the void volume of the Sephadex column, approximately the first 65 ml, are collected and pooled. The S-30 is stored as frozen spheres at $-196^o$ C and only thawed immediately before each assay. These spheres are prepared by dispensing the pooled excluded fractions, drop by drop, into liquid nitrogen with a long-tip Pasteur pipette. The activity of preparations stored in this manner is stable for at least six months.

The components of the cell-free wheat germ translation assay system are as follows: 24 mM Hepes (N-2-hydroxyethylpiperazine-N'-2 ethanesulfonic acid) (pH 7.6), 2 mM DTT (dithiothreitol), 1 mM ATP, 20 μM GTP, 8 mM creatine phosphate, 40 μg /ml creatine phosphokinase, 84 mM KCl, 2.5 mM Mg $acetate_2$, 20 μM each of the other unlabeled amino acids, and 4 μM [$^{14}$C]valine (260 Ci/mole). A 200-μl assay containing 40 μl of wheat germ S-30 is routinely used, and the incubations are performed for 2 hours at $25^o$ C. At the end of the incubation a 50 μl aliquot is removed, diluted into 2.0 ml of 5% TCA, and heated at 90° C for 10 minutes. The samples are then chilled at 4° C and the precipitates are collected on glass-fiber filters. The dried filters are placed in 5 ml of toluene-Spectrofluor and counted in a Beckman LS-250 liquid scintillation counter. The measurement of radioactivity incorporated into TCA-insoluble material is used as an indication of total mRNA activity. An additional 100-μl aliquot of the original reaction mixture is used for the determination of ovalbumin mRNA activity by a specific immunoprecipitation procedure [11] as follows: 25 μl of a monospecific ovalbumin antiserum and 5 μg of purified ovalbumin are added to the 100-μl aliquot after the addition of 30 μl of 5% Triton X-100 (v/v), 50 μl of 5% sodium deoxycholate (w/v), 10 mM valine, and 10 μl of 10 mM sodium phosphate (pH 7.0), containing 0.15M NaCl (to prevent nonspecific trapping). The immunoprecipitation reaction is allowed to proceed for 30 minutes at 25° C, and the resulting precipitate is collected on Millipore filters. The filters are rinsed with 15 ml of wash buffer (1% DOC, 1% Triton-X-100, 10 mM valine, 10 mM Na phosphate, 0.15M NaCl, pH 7.0), dried, and counted. Alternatively, the precipitate may be sedimented at 2000 x g for 10 minutes through a 1-ml sucrose cushion prepared in the wash buffer. The solution on top of the sucrose layer is removed and the surface is rinsed twice with wash buffer. After removal of the sucrose, the pellet is then rinsed two additional times with wash buffer, solubilized in 0.1 ml of NCS (Nuclear Chicago Co.), and counted. When a [$^{14}$C]ovalbumin internal standard is used to determine the efficiency of the immunoprecipitation reaction, only 85-90% of the TCA-precipitable radioactivity is found to be specifically precipitated. Thus, a correction of 10% is added to each value. In addition,the immunoprecipitation reaction is usually performed on the released, and presumably completed, peptide chains in the postribosomal supernatant (see below).

The specific activities of ovalbumin mRNA preparations are determined from the linear portions of the assay. Differing RNA inputs are used depending on the purity of the preparations tested. Thus, although 10-30 μg of the total RNA extract is assayed, only 0.4-2.0 μg of the final purified ovalbumin mRNA is tested.

Product analysis, without the use of immunoprecipitation, of the total peptides synthesized in the wheat germ assay in response to exogenous mRNA is performed by SDS gel electrophoresis of the released polypeptide chains. After a two-hour incubation, the ribosomes are removed by centrifugation at 105,000 x g for one hour at $4^o$ C. The postribosomal supernatant containing the released radioactive peptides is made 20 mM in $Na_2$ EDTA and incubated with (pancreatic) RNase A (20 μg/ml) for 15 minutes at $37^o$ C. Samples are then adjusted to 1% SDS and 0.01M DTT, heated at $90^o$ C for five minutes and dialyzed against 500 ml of 0.5% SDS, 1 mM DTT, 10 mM sodium phosphate, (pH 7.0) overnight. Analysis on 10% polyacrylamide gels containing 0.1% SDS is performed as described by Weber and Osborn [40]. The gels are cut into 2 mm slices using a Gilson gel fractionator, digested with 0.5 ml of 30% $H_2O_2$ at $60^o$ C overnight, and counted in a Spectrofluor-triton (2:1) scintillation cocktail.

When these techniques are used to assess the biologic purity of ovalbumin mRNA prepared as described in Section III, 92% of the total peptides synthesized in the wheat germ assay are specifically immunoprecipitable with the ovalbumin antiserum. Initially only 45% of the total mRNA activity is ovalbumin mRNA in the total RNA extract. A progressive increase in the percentage of ovalbumin mRNA is observed throughout the purification scheme. The specific activity of the purified ovalbumin mRNA determined in the wheat germ assay is also 140 times higher than the starting total RNA extract.

An additional proof of the purity of the final ovalbumin mRNA preparation is obtained by a careful product analysis of the total peptides synthesized in the wheat germ assay. A single peak is detected on SDS gels, which corresponds exactly to the ovalbumin synthesized in the wheat germ translation system, and then precipitated with a specific antibody to authentic ovalbumin run on a parallel gel. There are no detectable smaller molecular weight peptides, such as those observed when the peptides synthesized in response to a partially purified ovalbumin mRNA preparation obtained by Sepharose 4B chromatography are analyzed. Furthermore, more than 95% of the radioactivity present in this single band is immunoprecipitable with the monospecific ovalbumin antiserum.

### B. Physical Methods

Although the wheat germ translation assay is obviously of great utility for the determination of mRNA specific activities and for the detection of other contaminating mRNAs, other physical and chemical methods should also be used to assess mRNA purity.

Gel electrophoresis under denaturing conditions is a particularly sensitive technique.

In addition to the agarose gel electrophoresis method in 6M urea at pH 3.5 already described, an acrylamide gel electrophoresis that uses formamide as a denaturing agent has been used to analyze ovalbumin mRNA. The technique used in our laboratory is adapted from the method of Boedtker et al. [42]. Formamide (60 ml, Fisher, $A_{270} < 0.150$) is deionized by stirring for 1 hr. with 2 g of wet Dowex AG-501-X8, 20-50 mesh (BioRad). Dowex is removed by filtration through Whatman 42 filter paper and 0.136 g of $Na_2HPO_4$ and 0.33 g of $NaH_2PO_4$ are added to yield a concentration of 0.02M phosphate buffer (pH 7.4). Acrylamide (0.8 g) and bisacrylamide (0.12 g) (both BioRad electrophoresis grade) are dissolved in 19 ml of the buffered, deionized formamide and polymerized by the addition of 40 μl of TEMED (Eastman) and 0.32 ml of fresh ammonium persulfate, 75 μg/ml $H_2O$. Gels are usually allowed to polymerize overnight and preelectrophoresed at 5 mA/gel for 30 minutes at room temperature. The electrophoresis buffer is composed of 0.02 M phosphate buffer without formamide. The RNA samples are dissolved in 100 μl of buffered, deionized formamide to which 50 μl of a BPB-glycerine solution ( 5 ml of glycerine, 5 ml of deionized formamide, 60 μl of 1% BPB) is added. For such low-percentage gels dialysis tubing may be used to support the bottom of the gel during electrophoresis. After electrophoresis, gels may be fixed in 15% HAc to remove the formamide and scanned at 270 nm or stained with 1% methylene blue and scanned at 600 nm, as described. A detailed description of the formamide gel electrophoresis technique has recently been published [43]. Although this technique is applicable for the assessment of mRNA purity, caution should be used when it is utilized for molecular weight estimations (see Sec. V. A).

Analysis of mRNA preparations at each step of the described purification scheme by agarose gel electrophoresis reveals an increasing amount of a 21S RNA that correlates with the increased specific activity of ovalbumin mRNA. The final purified RNA migrates as a single band at 21S on both agarose-urea gels at pH 3.5 and in 99% formamide-containing polyacrylamide gels at neutral pH. It is completely free of contaminating 18S and 28S ribosomal RNA and DNA. Both of these techniques prevent RNA aggregation and can detect a minimum of 0.5 μg of RNA in a single band. Therefore, when 10 μg of RNA is analyzed, these techniques should be capable of detecting contamination at the 5% level or greater. Obviously, detection of the presence of minor contaminants which are heterogeneous in nature will be difficult.

The marked difference in ovalbumin mRNA migration during gel electrophoresis (21S) and during sedimentation on formamide-containing sucrose gradients (16S) also allows another means for detecting potential RNA contaminants. Presumably, this appreciable shift in sedimentation vs. migration relative to an 18S rRNA marker is a unique property of the (AU)-rich ovalbumin mRNA and will not occur for any contaminating (GC)-rich ribosomal RNA fragments (see Sec. V. A). Thus, when the purified ovalbumin mRNA is analyzed on formamide-containing sucrose gradients, it sediments as a sharp band at 16S, and no 21S RNA is detected by absorbance at 270 nm; moreover, all of the ovalbumin mRNA activity is again present in the 16S peak. Thus, sedimentation on formamide-containing sucrose gradients has also proved to be a useful analytic tool, in addition to the gel electrophoretic methods described, for the physical characterization of ovalbumin mRNA.

### C. Chemical Methods

Because ovalbumin mRNA cannot be labeled to high specific activity by incorporation of $^{32}P$ or tritiated nucleosides in vivo, chemical methods of labeling are used to generate the highly labeled ovalbumin mRNA necessary for hybridization experiments (see Sec. V. C) and fingerprinting analysis. Fingerprinting of partial RNase digests of $^{125}$I-labeled ovalbumin mRNA and of the $^{125}$I-labeled chick 18S and 28 rRNAs has proved to be an especially sensitive chemical method for determining ovalbumin mRNA purity. Both the 18S rRNA and fragments of 28S rRNA are potentially the principal non-mRNA contaminants in the 21S ovalbumin mRNA band. The methods used in our laboratory for the iodination of ovalbumin mRNA and fingerprinting are described below.

Iodination of RNA is performed by a modification of the method of Commerford et al [44]. The reaction mixture is composed of 0.1M NaAc (pH 5.0), 50 μM KI, 750 μM thallium chloride, 0.5 mg/ml of RNA, and 50 mCi/ml of carrier-free Na $^{125}$I in a total volume of 50 μl. After the reaction is allowed to proceed at 60° C for 15 minutes, 5 μl of 0.01M tyrosine is added to remove the unreacted iodide. Then 10 μl of 2.8M sodium phosphate, (pH 6.8), is added and the mixture is again incubated at 60° C for 15 minutes to destroy unstable intermediates. The entire reaction mixture is then cooled to room temperature and passed through a small Sephadex G-50 column (0.6 x 30 cm) in deionized water. Fractions in the void volume that contain radioactivity are pooled, and the labeled RNA is precipitated from alcohol at $-20^{\circ}$ C overnight after the addition of KCl to 0.5M. Precipitates are collected by centrifugation for 15 minutes at 10,000 g and redissolved in 100 μl of water. Then NaAc (pH 5.0) and $Na_2$ EDTA are then added to 0.01M each, and the $[^{125}I]$RNA is further purified by sedimentation on 0.3-1.0M sucrose gradients containing 0.01M NaAc (pH 5.0), 0.1M NaCl, and 1 mM EDTA for 16 hours at 35,000 rpm in a Beckman SW 40 rotor. Fractions of constant specific radioactivity are pooled, and the RNA thus labeled generally yields specific activities of $1\text{-}2 \times 10^7$ cpm/μg. The purified $^{125}I$-labeled RNA products have the same molecular weights as their unlabeled precursors, as demonstrated by electrophoresis on acid-urea agarose gels.

Degradation of the RNA during labeling is minimized by avoiding the alkaline pH conditions usually used during the destruction of unstable intermediates. In addition, tyrosine is used instead of the more reactive sodium sulfite to remove unreacted iodide. A final precaution involves purification and characterization of the $^{125}I$-labeled RNA. Occasionally a non-RNA particulate fraction is generated during the labeling procedure; this material is excluded from Sephadex G-25 and is precipitated by alcohol. This particulate material may represent a $^{125}I$-labeled thallium complex. It can be removed by sucrose gradient centrifugation and is usually detected at the bottom of the tube after centrifugation. If the final $^{125}I$-labeled RNA is not completely digested by pancreatic RNase (50 μg/ml for 30 minutes at 37° C), a further purification procedure should be utilized.

Generally, $^{125}I$-labeled RNAs appear to be of utility for analysis by finger-printing techniques. The labeling method has been shown to have little sensitivity to potential structure in single-stranded RNA molecules, yields stable oligonucleotide products in a reproducible manner, and does not change the specificity of several ribonucleases [45]. Here, $^{125}I$-labeled ovalbumin mRNA and chick 18S and 28S mRNAs are separately digested at 37° C for 45 minutes in 0.02M Tris-HCl (pH 7.4) - 2 mM EDTA with $T_1$ RNase (Sankyo Co. Ltd., Japan) or RNase A (Worthington, RAF) at an enzyme to substrate ratio of 1:20 (w/w). Each digest is fractionated by electrophoresis on Cellogel strips, followed by thin-layer homochromatography on the second dimension [46]. Cellogel strips, 3 x 95 cm (Reeve Angel Sci. Ltd., England), are first soaked in 7M urea-5% HAc - 3 mM EDTA, ad-adjusted to pH 3.5 with pyridine. Electrophoresis at 700 V is run in the same pH 3.5 buffer without EDTA until the pink dye marker (Acid Fuchsin) has migrated 45-50 cm.

Samples fractionated on Cellogel are transferred to 45 x 40 cm DEAE-cellulose (1:7.5, w/w) thin layer plates at a position 5 cm from the longer side of the plate. Fractionation in the second dimension is at 60$^{o}$ C for 7 hours, using different combinations of homomixture B with homomixture $C_5$ or homomixture $C_{15}$. Each homomixture is a 3% solution of torula yeast RNA (Sigma, Grade VI) in 7M urea. For homomixture "B", the yeast RNA is then precipitated with ethanol at room temperature to remove salt before being dissolved in the urea solution. For "C" homomixtures, the yeast RNA is hydrolyzed in 1N KOH at room temperature for 5 minutes (mixture $C_5$), or 15 minutes (mixture $C_{15}$), neutralized with concentrated HCl, and precipitated with ethanol at room temperature before being dissolved in the urea solution. A homomixture consisting of 25% mixture B and 75% mixture $C_5$ (v/v) is used to fractionate $T_1$ RNase digests, whereas a homomixture of 20% mixture B and

80% mixture $C_{15}$ (v/v) is used for pancreatic RNase digests.

After fractionation in the second dimension, the thin layer plates are air dried and autoradiographed using Dupont Cronex 4 X-ray film. Each nucleotide fingerprint map is done with 0.5-1 x $10^7$ cpm of [$^{125}$I]RNA radioactivity. The $T_1$ maps of mixtures of [$^{125}$I]mRNA with 18S [$^{125}$I]rRNA or 28S [$^{125}$I]rRNA contain 1 x $10^7$ cpm of each RNA species in approximately equal masses.

The analysis of each of these RNAs reveals the presence of a distinct fingerprint. The nucleotide fingerprint pattern is quite reproducible from different preparations, and a characteristic pattern of the large oligo-nucleotide spots is observed. Furthermore, each mixture map exhibits localized patterns of oligonucleotides that are characteristic of the individual RNA species. Thus, the absence of ribosomal RNA-specific patterns of large oligo-nucleotides for the fingerprint map of ovalbumin mRNA confirms that the mRNA is not appreciably contaminated with 18S or 28S ribosomal RNAs or their degradation products. This sensitive chemical method can, therefore, be used to complement the analysis performed by biologic and physical methods previously cited.

In the final section of this chapter, some of the additional methods used to characterize purified ovalbumin mRNA will be briefly discussed.

## V. MESSENGER RNA CHARACTERIZATION

### A. Molecular Weight Estimation

An accurate determination of the size of ovalbumin mRNA is a necessary prerequisite for any studies concerning mRNA transcription and processing. Unfortunately, radically different values are obtained when the molecular weight of ovalbumin mRNA is estimated by gel electrophoresis and sucrose gradient centrifugation. Thus, although ovalbumin mRNA is routinely observed as a 16S species in sucrose gradients under both nondenaturing [11] and denaturing (70% formamide) conditions [18], it migrates slower than 18S on both agarose-urea and formamide-polyacrylamide gels [18, 19]. The experimentally determined molecular weights range, therefore, from 520,000 or approximately 1600 nucleotides to 900,000, or approximately 2620 nucleotides. In order to resolve this discrepancy, two additional methods of molecular weight determination have been utilized.

First, the contour length of ovalbumin mRNA is directly measured by electron microscopy after spreading in a formamide-urea buffer by the method of Robberson et al [47]. In this technique, the RNA samples at 50 μg/ml in deionized water are diluted 50- to 100-fold and completely denatured by mixing into a solution of 4M urea dissolved in recrystallized formamide (final concentration of 80%) at 55° for 20 sec. Cytochrome c, a basic protein, is added to bind RNA. The sample is then layered onto a hypophase of distilled water, to form a surface film. The surface is touched by EM grids coated with a film of parlodion. The grids are stained with uranyl acetate, rotary-shadowed with Pt-Pd (80:20), and examined in a Philips 300 electron microscope. Photographs are taken on 35 mm film, and the image is enlarged by projection. The molecules are traced, and their contour lengths determined from the tracings with a "map measure." Magnification is calibrated with a germanium-coated replica with a diffraction grating of 21,600 lines/cm (Ladd).

When purified hen oviduct 18S rRNA and ovalbumin mRNA are thus examined under an electron microscope, both molecules are found to be completely unfolded. Ovalbumin mRNA has an average length of 0.5 ± 0.05 μm, which is slightly shorter than the purified 18S rRNA (0.55 μm). Since the average residue spacing is 2.65-

A/base [47], ovalbumin mRNA is thus composed of 1890 ± 190 nucleotide residues. If an average residue molecular weight of 343 (sodium salt) adjusted for the base composition of ovalbumin mRNA is used, a molecular weight of approximately 650,000 ±63,000 is obtained. The reproducibility of this technique is demonstrated by the essentially superimposable size distributions determined for two different preparations of purified ovalbumin mRNA.

The molecular weight of ovalbumin mRNA can also be determined by another independent method, which relies on the measurement of both the average length of the poly(A) tract and the poly(A) content in ovalbumin mRNA. Because of the difficulty in labeling ovalbumin mRNA and its poly(A) tail in vivo, both of these determinations utilized a specific and sensitive hybridization assay with [$^{3}$H]poly(U) [48]. In this method, [$^{3}$H]poly(U)of high specific activity (23 Ci/mole, Schwarz/Mann) is incubated with the poly(A)-containing mRNA under conditions of poly(U) excess, which favors the formation of a triple helix. A linear standard curve, in which unlabeled poly(A) at concentrations ranging from 0.05 to 0.4 μg is used, is run for each assay. Hybridization is performed in 0.45M NaCl, 45 mM Na citrate, 10 mM Tris-HCl (pH 7.2), containing 50% formamide at 36° C. RNA samples and [$^3$H]poly(U) dissolved in the hybridization buffer are sealed in disposable 50- or 100- μl capillaries and heat denatured at 62° C for 10 minutes. Hybridization is usually allowed to proceed overnight for convenience, even though poly(U-A) triple helix formation is extremely rapid. After incubation overnight, the samples are diluted into 1.0 ml of enzyme buffer containing 0.5M NaCl, 10 mM Tris (pH 7.2), 10 mM $MgCl_2$, 50 μg/ml pancreatic RNase II, and 20 μg/ml of DNase. The nonhybridized [$^3$H]poly(U) is digested for two hours at 30° C, and the samples are chilled in ice. After the addition of TCA to a final concentration of 10%, the hybrids are collected on nitrocellulose filters (Millipore, 0.45 μm, HAWP).

This sensitive hybridization assay may be used to localize poly(A)-containing mRNA on sucrose gradients, on agarose or polyacrylamide gels, or during Sepharose 4B chromatography. A close correspondence is usually observed between the detection of mRNA activity in a cell-free translation assay and the presence of poly(A)-containing RNA determined with [$^3$H]poly(U). By this method, purified ovalbumin mRNA is found to contain approximately 4.25% poly(A).

The average length of poly(A) in purified ovalbumin mRNA may also be determined by using the [$^3$H]-

poly(U) hybridization probe. Here, 20 μg of purified ovalbumin mRNA is dissolved in 50 μl of 0.02M Tris-HCl (pH 7.6), containing 2 mM EDTA and 0.3M NaCl. Then, 3 units of pancreatic ribonuclease A and one unit of ribonuclease T1 are added, and the solution is incubated at 37° C for 1 hour. The digest is subjected to electrophoresis on 12% polyacrylamide gels in a Tris-EDTA buffer at pH 8.6 [49]. The gels are sectioned into 1 mm slices after electrophoresis, and each gel slice is homogenized in 0.1 ml of 0.5M NaCl with a Teflon pestle. The homogenate is centrifuged at 12,000 x g for 15 minutes after standing at 2° C overnight and the supernatants are used for the [$^3$H]-poly(U) hybridization assay. The slices that yield hybridizable radioactivity are assumed to contain the poly(A) chains. The migration of poly(A) is then measured, and its average chain length is determined by comparing the migration of the peak of hybridizable radioactivity to that of standards of poly(A) of known chain lengths (Miles). Since the migration of poly(A) on nondenaturing gels is considerably retarded in comparison to 5S and 4S RNAs, the use of poly(A) standards is necessary for the accurate determination of the poly(A) chain lengths. For example, a 5S RNA standard of approximately 120 nucleotides in length actually migrates at a rate faster than that of a poly(A) standard of 90 nucleotides.

The poly(A) in a purified ovalbumin mRNA digest migrates as a broad spectrum between standard poly(A) of 90 and 45 residues, with a number-average chain length of 62 adenylate residues. A rough estimation of the molecular weight of ovalbumin mRNA is therefore obtained in the following manner: Since 62 nucleotides constitute approximately 4.20% of the entire molecule, the purified ovalbumin mRNA should be composed of approximately 1476 nucleotides or 510,000 daltons. This value is obviously an approximation, since a braod distribution of poly(A) sizes is found after nuclease digestion. But, this molecular weight estimation is in close agreement with the value originally obtained from the sucrose gradient centrifugation method. The accuracy of the molecular weight analysis by electron microscopy is considerably superior to both the poly(A) analysis and the sucrose gradient centrifugation method. Moreover, regardless of their accuracies, all three methods yield molecular weight values that are markedly less than those obtained by gel electrophoresis.

The overestimation of molecular weights from

data obtained by gel electrophoresis is by no means unique to ovalbumin mRNA. This discrepancy is also observed for other "low GC-content" RNAs, such as mitochondrial rRNA [47] and hemoglobin mRNA [33], when their molecular weights are calculated using higher GC-content rRNA standards. The migration of RNAs during gel electrophoresis is a reflection of both RNA chain length and secondary structure. Thus, even under the denaturing conditions generally used, secondary structure may still be present in 28S and 18S ribosomal RNAs with their higher GC content. A striking example is the inability of 80% formamide and 4M urea to completely disrupt the secondary structure in 28S rRNA and the visualization of these structures in the electron microscope [50]. Therefore, until appropriate molecular weight standards are available, care should be taken to use several independent methods of molecular weight analysis for the determination of the molecular weight of mRNAs. Only then can the length of any noncoding portions in the mRNA be accurately determined.

Using the value obtained by length measurement in the electron microscope as our best present estimate, ovalbumin mRNA contains approximately 600 additional noncoding bases, exclusive of the poly(A) region.

since the coding portion of the molecule should contain approximately 1200 bases (387 amino acids), ovalbumin mRNA is, therefore, 50% longer than necessary. The function of these noncoding portions of the molecule is unfortunately only a matter of speculation at the present time. These sequences may be important in the maintenance of the helical structure observed in purified ovalbumin mRNA. Approximately 25% helical structure has been observed in ovalbumin mRNA when ethidium bromide probing is used (N. T. Van, unpublished observations). In addition, these regions may also be involved in the stabilization of ovalbumin mRNA, accounting for its unusually long half-life of 24 to 48 hr. [26]. The availability of large quantities of purified ovalbumin mRNA and intact, $^{125}$I-labeled ovalbumin mRNA of high specific activity should greatly facilitate future studies designed to elucidate both the location and function of these sequences within the ovalbumin mRNA molecule.

### B. Hybridization Analysis*

The determination of the complexity of a purified mRNA may be accomplished by a careful analysis of its kinetics of hybridization under conditions of DNA

---

*(See Chap. 2.)

excess. Because of the complexity of eukaryotic DNA, these experiments require the use of nucleic acids of high specific activity and sequence ratios of at least $10^6$ to 1. A complementary [$^3$H]DNA copy of ovalbumin mRNA synthesized by means of a viral RNA-directed DNA polymerase [12] has been used to determine if the ovalbumin gene is amplified or present only once in the chick genome. Results from several laboratories [12, 53] have substantiated the possibility that ovalbumin mRNA is transcribed from unique sequence DNA. But, the complementary DNA used in these studies is primarily representative of only 10% of the ovalbumin mRNA molecule and is presumably a copy of its 3'-end (S. Harris and J. Monahan, personal communication). This analysis does not, therefore, reveal the complexity of the entire molecule. The sequence complexity of the entire molecule can, however, be analyzed using ovalbumin [$^{125}$I]mRNA under conditions of DNA excess.

These are two potential limitations in this type of analysis. An accurate determination of the kinetic complexity of an unknown mRNA requires a comparison with the reassociation of a DNA or RNA of known complexity. Differences in fragment size, GC-content, and rates of DNA vs. RNA hybridization may all affect these relative rates of hybridization. Furthermore, DNA-excess hybridization reactions,even when high DNA/RNA complementary sequence ratios are used, rarely go to completion. The maximal extent of hybridization observed for mRNA preparations is usually only 50%. Since this value may in part reflect mRNA degradation during long-term incubations, our hybridizations are performed in 50% formamide at reduced temperatures ($41^{\circ}$ C instead of $70^{\circ}$ C). Under these conditions there is no loss of TCA-precipitable [$^{125}$I]mRNA radioactivity during eight days of incubation. A critical factor in all of these experiments is the use of rigorously purified nucleic acids. Care must be taken to avoid RNase contamination and the presence of heavy metal ions. The latter may be avoided by the use of EDTA in the hybridization buffer and by passage of all buffers and nucleic acids through Chelex resin (BioRad) before use.

Our protocol is as follows: Total chick liver DNA is prepared by conventional methods [51]. All DNA preparations used are essentially free of RNA and protein as determined by diphenylamine analysis and thermal denaturation profiles. The final DNA solution is adjusted to approximately 1 mg/ml in

0.2M NaCl and sheared twice in a French press to yield DNA fragments of approximately 400 base pairs in length. Pronase is then added to 50 μg/ml and the solution is incubated for 1 hour at 37° C to remove any possible trace contamination of ribonuclease. Then, SDS is added to 0.5% and the solution is vigorously shaken for five minutes with an equal volume of a pH 8.0, phenol-SDS buffer identical to that used for total nucleic acid extraction. The mixture is then chilled for 20 minutes and centrifuged for 10 minutes at 10,000 g. The aqueous phase is extracted two more times. The sheared DNA in the aqueous phase is finally precipitated with two volumes of ethanol at -20° C. The precipitate is collected by centrifugation at 10,000 g, dried under gentle vacuum, and redissolved in a small volume of the hybridization buffer - 0.01M N-Tris-methyl-2-aminoethane sulfonic acid (TES), containing 1 mM $Na_2$ EDTA, 0.75M NaCl, and 50% formamide, pH 7.0 - to yield a DNA/RNA ratio of $3 \times 10^6$ to 1. At this ratio, each 100-μl aliquot contains at least $5 \times 10^3$ cpm. Aliquots of 100 μl are sealed in capillary tubes and heated at 70° C for 10 minutes to denature the DNA. Hybridization is allowed to proceed at 41° C for various lengths of time, up to eight days. The incubation is terminated by a 50-

fold dilution into 4.9 ml of ice-cold 0.24M Na phosphate (pH 6.8) and the tubes are frozen at -20° C until assayed for hybrid formation. All the reaction time points are assayed simultaneously in the following manner: the tubes are thawed and their contents dispensed into two 2-ml and one 0.5-ml aliquots; one 2-ml portion is treated for one hour at room temperature with a mixture of pancreatic ribonuclease(A) (50 μg/ml) and ribonuclease ($T_1$) (50 units/ml) and precipitated with cold 100% TCA added to a final concentration of 10%; the nucleic acid in the other 2-ml portion is precipitated immediately with cold TCA without prior ribonuclease treatment. The TCA-insoluble radioactivity is collected on glass filters and washed with cold 10% TCA. Each filter is then incubated in 1 ml of NCS for two hours at 37° C to solubilize filter-bound radioactivity. Then 10 ml of spectrofluor-toluene scintillation fluid is added and the sample is allowed to stand for 16 hours in the dark. The percent hybridization is determined by calculating the ratio of the ribonuclease-resistant radioactivity to the total TCA-precipitable radioactivity. The values are corrected for a control value obtained by determining the ribonuclease resistance of the [$^{125}$I]mRNA in the reaction mixture in which the DNA is not denatured. The remaining 0.5-ml aliquots of the assay mixture are used to determine the extent of DNA-DNA reassociation by hydroxyapatite chromatography [12].

When such an experiment is performed at a DNA/RNA ratio of 3 x $10^6$, only a single transition in the hybridization curve is detected, at an observed $C_ot_{1/2}$ value of 550. Under these conditions the observed $C_ot_{1/2}$ value of total chick DNA is 170. A background value of ribonuclease resistance of 7% is also obtained, which may reflect intramolecular foldback in the ovalbumin mRNA molecule. This value is subtracted from the percent hybridization. The maximum extent of hybridization observed is only 23% of the input [$^{125}$I]mRNA. This result is not totally unexpected since at a complementary sequence of only 3 [i.e., 1890 nucleotides(molecular weight of ovalbumin mRNA) ÷ 2 x $10^9$ nucleotides (molecular weight of chick DNA) = $10^6$; therefore, at a DNA/RNA ratio of 3 x $10^6$, the complementary sequence ratio = 3] the theoretical maximal percent hybridization under conditions of DNA excess may be two to four times less than comparable DNA reassociation rates under similar hybridization conditions (E. Davidson, personal communication). Thus, considering these factors and the lack of any major early transition in the hybridization curve, it is not unreasonable to conclude that the majority of the mRNA molecule is transcribed from unique sequence DNA in the chick genome. The lack of a major repetitive component in highly purified ovalbumin mRNA is consistent with data obtained for other eukaryotic mRNAs [52].

These hybridization experiments are necessary prerequisites for future experiments in which attempts will be made to isolate the gene for ovalbumin mRNA. Prior knowledge of the sequence complexity of ovalbumin mRNA and of the sequence organization within chick DNA is required for the successful design of these gene isolation experiments. The ultimate elucidation of steroid hormone action at the molecular level may require the reconstitution of steroid hormone-receptor complexes with specific gene sequences in vitro. Thus, the isolation and characterization of large quantities of purified ovalbumin mRNA by the techniques summarized in this chapter provides an important first step in attaining this goal.

## ACKNOWLEDGMENTS

We thank Drs. C. Liarkas and Y. Choi for the inclusion of their modified Brownlee and Sanger homochromatography procedure prior to publication and to Dr. D. Robberson for providing the details of the electron microscopic procedure and for several helpful suggestions concerning methods of molecular weight analysis.

## REFERENCES

1. J. Stavnezer and R. C. Huang, Nature (New Biol.), 230, 172 (1971).
2. R. E. Lockard and J. B. Lingrel, Biochem. Biophys. Res. Comm., 37, 204 (1969).
3. G. R. Molloy, M. B. Sporn, D. E. Kelley, and R. P. Perry, Biochemistry, II, 3256 (1972).
4. H. Aviv and P. Leder, Proc. Natl. Acad. Sci. U. S. A., 69, 1408 (1972).
5. G. Brawerman, I. Mendecki and S. Y. Lee, Biochemistry, 11, 637 (1972).
6. B. W. O'Malley, W. L McGuire, P. O. Kohler and S. G. Korenman, Recent Prog. Horm. Res., 25, 105 (1969).
7. P. O. Kohler, P. Grimley and B. M. O'Malley, Science, 160, 86 (1968).
8. P. E. Kohler, P. Grimley and B. W. O'Malley, J. Cell Biol., 40, 8 (1969).

9. T. Oka and R. T. Schimke, J. Cell Biol., 41, 816 (1969).

10. R. D. Palmiter and L. T. Smith, Nature (New Biol., 40, 8 (1969).

11. A. R. Means, J. P. Comstock, G. C. Rosenfeld and B. W. O'Malley, Proc. Natl. Acad. Sci. U. S. A., 69, 1146 (1972).

12. J. M. Rosen, S. E. Harris, G. C. Rosenfeld, C. Liarakas, B. W. O'Malley and A. R. Means, Cell. Diff., 3, 103 (1974).

13. B. W. O'Malley and A. R. Means, Science, 183, 610 (1974).

14. J. M. Rosen, S. L. C. Woo, and J. P. Comstock, Biochemistry, submitted for publication (1975).

15. R. P. Perry, J. LaTorre, D. E. Kelley, and J. Greenberg, Biochim. Biophys. Acta, 262, 220 (1972).

16. R. E. Rhoads, G. S. McKnight and R. T. Schimke, J. Biol. Chem., 248, 2031 (1973).

17. J. W. Sedat and R. L Sinsheimer, Cold Spring Harbor Symp. Quant. Biol., 35, 163 (1970).

18. J. M. Rosen, S. L. C. Woo, J. W. Holder, A. R. Means and B. W. O'Malley, Biochemistry, 14, 69 (1975).

19. M. E. Haines, N. H. Carey and R. D. Palmiter, Eur. J. Biochem., 43, 549 (1974).

20. R. D. Palmiter, J. Biol. Chem., 248, 2095 (1973).

21. R. D. Palmiter, Biochemistry, 13, 3606 (1974).

22. R. Palacios, D. Sullivan, N. M. Summers, M. L. Kiely and R. T. Schimke, J. Biol. Chem., 248, 540 (1973).

23. D. J. Shapiro, J. M. Taylor, G. S. McKnight, R. Palacios, C. Gonzalez, M. L. Kiely and R. T. Schimke, J. Biol. Chem., 249, 3665 (1972).

24. R. Palacios, R. D. Palmiter, and R. T. Schimke, J. Biol. Chem., 247, 2316 (1972).

25. R. D. Palmiter, R. Palacios and R. T. Schimke, J. Biol. Chem., 247, 3296 (1972).

26. S. E. Harris, J. M. Rosen, A. R. Means and B. W. O'Malley, Biochemistry, 14, 2072 (1975).

27. R. F. Cox, M. E. Haines and J. S. Emtage, Eur. J. Biochem., 49, 225 (1974).

28. G. S. McKnight and R. T. Schimke, Proc. Natl. Acad. Sci. U. S. A., 71, 4327 (1974).

29. R. Williamson, M. Morrison, and J. Paul, Biochem. Biophys. Res. Commun., 40, 740 (1970).

30. Y. Suzuki, L. P. Gage and D. D. Brown, J. Mol. Biol., 70, 637 (1972).

31. J. M. Rosen and B. W. O'Malley in Biochemical Action of Hormones, G. Litwack, ed. Vol. III, Acadmic Press, New York, 1975.

32. I. Schechter, Biochem. Biophys. Res. Commun., 57, 857 (1974).

33. R. Williamson, M. Morrison, G. Lanyon, R. Eason and J. Paul, Biochemistry, 10, 3014 (1971).

34. S. L. C. Woo, J. M. Rosen, C. D. Liarkos, D. Robberson, J. Busch, A. R. Means, and B. W. O'Malley, J. Biol. Chem., submitted for publication (1975).

35. P. A. Weil and A. Hampel, Biochemistry, 12, 4361 (1973).

36. U. Lindberg and T. Persson, Eur. J. Biochem., 31, 246 (1972).

37. J. N. Mansbridge, J. A. Crossley, W. G. Lanyon and R. Williamson, Eur. J. Biochem., 44, 261 (1974).

38. B. E. Roberts and B. M. Paterson, Proc. Natl. Acad. Sci. U. S. A., 70, 2330 (1973).

39. A. Marcus, D. P. Weeks, J. Leis and E. B. Keller, Proc. Natl. Acad. Sci. U. S. A., 67, 1681(1970).

40. K. Weber, and M. Osborn, J. Biol. Chem., 244, 4406 (1969).

41. C. L. Privero, H. Aviv, B. M. Paterson, B. E. Roberts, S. Rosenblatt, M. Revel and E. Vinocour, Proc. Natl. Acad. Sci. U. S. A., 71, 302 (1974).

42. H. Boedtker, R. B. Crkvenjakov, K. F. Dewey and K. Lanks, Biochemistry, 12, 4356 (1973).

43. J. C. Pinder, D. Z. Staynov and W. B. Gratzer, Biochemistry, 13, 5373 (1974).

44. S. L. Commerford, Biochemistry, 10, 1993, (1971)

45. H. D. Robertson, E. Dickson, P. Model and W. Prensky, Proc. Natl. Acad. Sci. U. S. A., 70, 3260 (1973).

46. G. G. Brownlee,and F. Sanger, Eur. J. Biochem., II, 395 (1969).

47. D. Robberson, Y. Aloni, G. Attardi and N. Davidson, J. Mol. Biol., 60, 473 (1971).

48. G. Gillespie, J. Marshall and R. C. Gallo, Nature(New Biol.), 236, 227 (1972).

49. U. Loening, Biochem. J., 113, 131 (1969).

50. P. Wellauer, I. Dawid, D. E. Kelley and R. P. Perry, J. Mol. Biol., 89, 397(1974)

51. J. Marmur, J. Mol. Biol., 3, 208 (1961).

52. R. B. Goldberg, G. A. Galau, R. J. Britten and E. H. Davidson, Proc. Natl. Acad. Sci. U. S. A., 70, 3516 (1973)

53. D. Sullivan, R. Palacios, J. Stavnezer, J. M. Taylor, A. J. Faras, M. L. Kiely, N. M. Summers, J. M. Bishop and R. T. Schimke, J. Biol. Chem., 248, 7530 (1973).

# AUTHOR INDEX

Numbers in brackets are reference numbers and indicate that an author's work is referred to although his name is not cited in the text. Underlined numbers give the page on which the complete reference is listed.

B

D

H

I



J



K

N

## SUBJECT INDEX

X



Z